Population and Commu

DYNAMICS OF NUTRIENT CYCLING AND FOOD WEBS

Population and Community Biology Series

Principal Editor

M. B. Usher
Chief Scientific Advisor, Nature Conservancy Council for Scotland, UK and Reader, Department of Biology, University of York, UK

Editors

M. L. Rosenzweig
Professor, Department of Ecology and Evolutionary Biology, University of Arizona, USA

R. L. Kitching
Professor, Department of Ecosystem Management, University of New England, Australia

The study of both populations and communities is central to the science of ecology. This series of books explores many facets of population biology and the processes that determine the structure and dynamics of communities. Although individual authors are given freedom to develop their subjects in their own way these books are scientifically rigorous and a quantitative approach to analysing population and community phenomena is often used.

Already published

1. **Population Dynamics of Infectious Diseases**
   Theory and applications
   Edited by R. M. Anderson

2. **Food Webs**
   Stuart L. Pimm

3. **Predation**
   Robert J. Taylor

4. **The Statistics of Natural Selection**
   Bryan F. J. Manly

5. **Multivariate Analysis of Ecological Communities**
   P. Digby and R. Kempton

6. **Competition**
   Paul A. Keddy

7. **Stage-Structured Populations**
   Sampling, analysis and simulation
   Bryan F. J. Manly

8. **Habitat Structure**
   The physical arrangement of objects in space
   Edited by S. S. Bell, E. D. McCoy and H. R. Mushinsky

DYNAMICS OF NUTRIENT CYCLING AND FOOD WEBS

D. L. DeAngelis

Environmental Sciences Division
Oak Ridge National Laboratory
Tennessee, USA

CHAPMAN & HALL
London · New York · Tokyo · Melbourne · Madras

Published by Chapman & Hall, 2–6 Boundary Row, London SE1 8HN

Chapman & Hall, 2–6 Boundary Row, London SE1 8HN, UK

Chapman & Hall, 29 West 35th Street, New York NY10001, USA

Chapman & Hall Japan, Thomson Publishing Japan, Hirakawacho Nemoto Building, 7F, 1-7-11 Hirakawa-cho, Chiyoda-ku, Tokyo 102, Japan

Chapman & Hall Australia, Thomas Nelson Australia, 102 Dodds Street, South Melbourne, Victoria 3205, Australia

Chapman & Hall India, R. Seshadri, 32 Second Main Road, CIT East, Madras 600 035, India

First edition 1992
© 1992 D. L. De Angelis

Typset in 10/12pt Times by
Interprint Ltd, Malta
Printed in Great Britain by
St. Edmundsbury Press, Bury St. Edmunds, Suffolk

ISBN 0 412 29830 9 (HB) 0 412 29840 6 (PB)

A catalogue record for this book is available from the British Library

Library of Congress Cataloging-in-Publication data enclosed
DeAngelis, D. L. (Donald Lee), 1944
      Dynamics of nutrient cycling and food webs / D. L. DeAngelis.—1st ed.
      p. cm.—(Population and community biology series)
      Includes bibliographical references and index.
      ISBN 0–412–29830–9 (HB). —ISBN 0–412–29840–6 (PB)
      1. Biogeochemical cycles. 2. Food chains (Ecology) I. Title. II. Series.
IN PROCESS (ONLINE)
574.5′3—dc20   91–3275
                                                                                                              CIP

Contents

Acknowledgements

Acknowledgments are made to the following for permission to reproduce or modify Tables and figures: Figure 1.2 from *Energy Flow in Biology* by Morowitz, H. J. (published 1968 by Academic Press); Figure 1.4 from *Estuarine Ecology* by Day, Jr., J. W., Hall, C. A. S., Kemp, W. M. and Yanez-Arancibia, A. (Copyright 1989 by John Wiley & Sons, Inc.); Figure 2.3 from article by Canfield, D. E., Green, W. J., Gardner, T. J. and Ferdelman, T., *Archiv fur Hydrobiologie* **100** 501–19 [published 1984 by Schweizerbart'sche Verlagsbuchhandlung, E. (Nagele u. Obermiller)]; Figure 2.5 from article by Kirchner, W. B. and Dillon, P. J., Water Resources Research **11** 182–3 (published 1975 by the American Geophysical Union); Figures 2.8 and 2.9 from article by Ahlgren, I., *Archiv fur Hydrobiologie* **89** 17–32 [published 1980 by Schweizerbart'sche Verlagsbuchhandlung, E. (Nagele u. Obermiller)]; Figure 3.1 from article by Switzer, G. L. and Nelson, L. E., *Soil Science Society of America Proceedings* **36** 143–7 (published 1972 by the Soil Science Society of America, Inc.); Figure 3.5 from article by Welch, E. B., Hendrey, G. R. and Stoll, R. K., Oikos (published 1975 by Munksgaard International Booksellers and Publishers); Figure 4.1 from drawn using data from Table 3 of article by Clarkson, D. T., *Journal of Ecology* **55** 111–18 (published by the British Ecological Society); Figures 4.2 and 4.3 from an article by Droop, M. R., *Journal of Phycology* **9** 264–72 (published by the Phycological Society of America); Figure 4.10 from article by M. Gatto and S. Rinaldi, Vegetatio **69** 213–22 (published 1987 by Dr. W. Junk); Figures 5.9 and 5.10 from *The herbivore-based trophic system*, by Batzli, G. O., White, R. G., MacLean, S. F., Pitelka, F. A. and Collier, B. D. in *An Arctic Ecosystem* (published 1980 by Dowden, Hutchinson, and Ross, Publishers); Figure 6.1 from an article by Sterner, R. W., *Science* **231** 605–7 (copyright 1986 by the American Association for the Advancement of Science); Figure 6.2 from an article by Carpenter S. R. and Kitchell, J. F., *The American Naturalist* **124** 159–72 (copyright 1984 by the University of Chicago Press); Figures 6.5, 6.6, and 6.7 from an article by Kiefer, D. A. and Atkinson, C. A., *Journal of Marine Research* **42** 655–75 (published 1984 by the Sears Foundation for Marine Research; Figure 7.3 from an article by Ruess R.W. and McNaughton, S. J. Oikos **49** 101–10 (published 1987 by Munksgaard International Booksellers and Publishers); Figure 7.6 from an article by Parnas, H., *Soil Biology and Biochemistry* **7** 161–9, *Model for decomposition of organic material by microorganisms* (copyright 1975 by Pergamon Press, Inc.); Figure 7.7 from an article by Ingham, R. E.,

Trofymow, J. A., Ingham, E. R. and Coleman, D. C., *Ecological Monographs* **55** 119–40 (published 1985 by the Ecological Society of America); Figure 8.2 from an article by McQueen, D. J., Post, J. R. and Mills, E. L., *Canadian Journal of Fisheries and Aquatic Sciences* (published 1986 by the Department of Fisheries and Oceans, Canada); Figures 8.5 and 8.6 from an article by Finn, J. T., *Journal of Theoretical Biology* **99** 479–89 [published 1982 by Academic Press, Ltd. (London)]; Figures 9.3, 9.4, and 9.5 from an article by Armstrong, R. A., *Ecology* **60** 76–84 (published 1979 by the Ecological Society of America); Figures 9.6 and 9.7 from an article by Olson, P. and Willen, E., *Archiv fur Hydrobiologie* **89** 171–88 [published 1980 by Schweizerbart'sche Verlagsbuchhandlung, E. (Nagele u. Obermiller)]; Figures 9.10 and 9.12 from *Resource Competition and Community Structure*, by Tilman, D. (published 1982 by Princeton University Press); Figures 10.4, 10.5 and 10.6 from an article by Harrison G. W. and Fekete, S., *Ecological Modelling* **10** 227–41 (published 1980 by Elsevier Science Publishers); Figure 10.7 from an article by DeAngelis, D. L., Waterhouse, J. C., Post, W. M. and O'Neill, R. V., *Ecological Modelling* **29** 399–419 (published 1985 by Elsevier Science Publishers); Figures 10.8, 10.9, and 10.10 from an article by Kemp, W. M. and Mitsch, W. J., *Ecological Modelling* **7** 201–22, (published by Elsevier Science Publishers); Figure 10.11 from an article by Hessen D. O. and Nilssen, J. P., *Archiv fur Hydrobiologie* **105** 273–84 [published by Schweizerbart'sche Verlagsbuchhandlung, E. (Nagele u. Obermiller)]; Figures 11.2 and 11.3 from an article by Newbold, J. D., O'Neill, R. V., Elwood, J. W. and Van Winkle, W., *The American Naturalist* **120** 628–52 (copyright 1982 by the University of Chicago Press); Figure 11.4 from *The Mathematics of Diffusion*, by Crank, J. (published 1979 by Oxford University Press); Figure 12.2 from Ocean Tracers Laboratory Technical Report #4, Department of Geological and Geophysical Sciences, Princeton University, by Toggweiler, J. R., Sarmiento, J. L., Najjar, R. and Papademetriou, D.

Preface

In all fields of science today, data are collected and theories are developed and published faster than scientists can keep up with, let alone thoroughly digest. In ecology the fact that practitioners tend to be divided between such subdisciplines as aquatic and terrestrial ecology, as well as between population, community, and ecosystem ecology, makes it even harder for them to keep up with all relevant research. Ecologists specializing in one subdiscipline are not always aware of progress in another subdiscipline that relates to their own. Syntheses are frequently needed that pull together large bodies of information and organize them in ways that makes them more coherent, and thus more understandable. I have tried to perform this task of integration for the subject area that encompasses the interrelationships between the dynamics of ecological food webs and the cycling of nutrients. I believe this area cuts across many of the subdisciplines of ecology and is pivotal to our progress in understanding ecosystems and in dealing with human impacts on the environment. Many current ecological problems involve human disturbances of both food webs and the nutrients that cycle through them.

Little progress can be made towards elucidating the complex feedback relations inherent in the study of nutrient cycles in ecological systems without the tools of mathematics and computer modelling. These tools are therefore liberally used throughout the book. Despite this, a knowledge of mathematical concepts such as stability and eigenvalues is not assumed beforehand. This book should be suitable to graduate students in ecology who have some acquaintance with elementary differential equations.

The book begins with elementary models of material flux in physical systems, such as single-compartment models of lakes. From these, most of the important analytical approaches needed for studying more complex systems are developed. Autotrophs, heterotrophs, and decomposers are added a component at a time, and each new aspect of the dynamics is described as the material cycles are integrated into the living components of more complex food webs. The special topics of temporal variability of food webs and the effects of spatial heterogeneity are discussed in separate chapters, as both of these subject areas are of key importance to the current understanding of food webs.

My efforts to synthesize ideas on nutrient and cycling and food webs owe much, directly and indirectly, to interactions over the years with staff members at Oak Ridge National Laboratory, many of whom have commented on portions of the book, including Bartell, S. M., Elwood, J. W., Emanuel, W. R., Gardner, R. H., Huston, M. A., Luxmoore, P. J., O'Neill, R. V., Mulholland, P. J., Palumbo, V., Peng, T.-H. and Post, W. M. I am indebted to Usher, M. B. and Rosenzweig, M. L. for their encouragement of this book from the beginning and, in particular, to Professor Usher for his extensive criticism of early drafts of the manuscript. The figures were ably drawn by the graphics staff, including Adams, R. R., Booker, R. E., Holbrook, J. E., Rich, D. L. and Williamson, M. This work was supported by the National Science Foundations Ecosystems Studies Program under Interagency Agreement 40–689–78 with the US Department of Energy under contract DE-ACOS-840R21400 with Martin Marietta Energy Systems, Inc. Environmental Sciences Division Publ. No. 3719.

Foreword

Like everyone else, ecologists have difficulties in seeing the forest for the trees; for ecologists, the consequences of this short-sightedness are more severe. It is easy for us to see the trees and the birds and the butterflies around them, and most of what we do as ecologists involves direct observation and simple experimentation on such conspicuous species. The emphasis is on what is easy to observe, so we study the ecology of a very small number of species, for a few years, over a small area. What is important – 'the forest' – is rather different. The five billion or so humans on this planet are changing its climate, dumping vast amounts of pollutants into the air, the rivers, lakes and oceans, and felling large parts of the native forest. Certainly we lose the 'trees' – the individual species, some of economic importance (tropical plants are the source of many major drugs in the fight against cancer, for example), and some of intrinsic beauty or interest (who would want to tell their children that the rhinos, elephants and tigers of their story books were extinct?). But we also lose the 'forest', the complex of many species that provide us essential ecosystem services.

The problem is how do we study the 'forest'. That is, how do we comprehend ecological processes on the very large spatial and temporal scales that are required? There are two key concepts. The first addresses a problem of another biological metaphor: how to mix apples and oranges. How do we find common currencies that will allow not only the study of many species of plants, but also the myriad of animal species that depend on them, how all these species are affected by the physical environment, and how the species in turn affect the soil, the atmosphere and their adjacent rivers and lakes? The common currencies are the flows of essential nutrients – the flows of water, carbon, nitrogen and phosphorus, for example. The second key concept is the food web – the diagram that depicts the complexity of pathways these essential nutrients take into, through and out of the ecosystems. Common sense tells us that how ecosystems function will depend on the nature and complexity of the flows through them. DeAngelis's book is so important because for the first time, he treats these two concepts in a synthetic way. In doing so, he provides us with a major summary of ecosystem ecology.

The applied problems that appear throughout this book span an extraordinary range of scales. There are the issues of local pollution; for instance, in aquatic ecosystems, phosphorus is often in short supply naturally, but is one of the important ingredients in our industrial and agricultural waste and can have a major impact as a consequence. Nutrient flows are often radically altered when we clearcut forests, illustrating vividly why we must understand the roles of plants, animals and soils. There are the global issues of climate change caused largely by increasing emissions of carbon dioxide. Where the carbon in fossil fuels goes when we burn it is not merely into the atmosphere, because some of it then goes back into plants and the rest of the food web. We must understand how much goes where and how fast it gets there if we are to predict what the long-term changes will be.

Important conceptual issues are also developed in the text. Throughout is a discussion of resilience: the rate at which things change. This is a central issue, for we need to know how quickly we will observe changes and how quickly ecosystems will recover if we give them the chance. Not all ecosystems are the same in this respect, and cross-ecosystem comparisons are another feature of this book. DeAngelis asks many interesting questions. How does web structure alter resilience? What happens when we add or subtract a trophic level, for example, by excluding grazers from ecosystems? Are the effects of species sometimes greater than we might expect on the basis of their biomass alone? In short, this book is also a major investigation into the nature of ecosystem processes and provides both an up-to-date summary of the state of ecosystem ecology and a clear statement of how the field should develop in the future.

Ecologists have neglected ecosystem ecology because the problems the science poses are applied ones. Large-scale processes operating over long time periods make conventional experiments difficult or impossible and how can one develop a theoretical understanding of the so very complex ecosystem processes? Ecosystem ecology is intrinsically a very difficult science. Has this book overcome these difficulties and if so how?

There has been a tradition in ecosystem ecology of building complicated simulation models. Such models are often so complex that their workings are only comprehensible to those that build them, and their results are likely to be understood only with an effort that would rival that required to understand the ecosystems they are supposed to mimic. This means that ecosystem ecology can be completely inpenetrable to those from the outside. To me, the appearance of 'emergent properties' from models too complex for us to develop any insight into that emergence smacks of mysticism. In addition, my experience shows me that it is very easy to build models but very hard to build models that capture enough of the essence of the real world to be useful. What then is the purpose of building ecosystem models if we cannot test them; how do we know the models are not purely theoretical edifices? Although the models are the central core of this book, throughout there is constant checking of the models against what we know about ecosystems from observations, the intuitions of field ecologists and the occasional experiment. This is not theory in a vacuum.

Perhaps what impresses me most about this book is its clear emphatic statement of how we should study ecosystems. DeAngelis does not hide behind incomprehensible models. He starts off with simple models, carefully explained, and progresses gently towards ever greater complexity, developing the insights needed to understand this complexity along the way. Mathematical models are never easy, but their clear formulation and development and the special care taken to ensure that all the models use the same sensible notation makes the effort less daunting. Our students finally have a textbook on ecosystem ecology, future generations have a role model for lucid analysis, and ecologists have a superb synthesis of the science's most difficult arena. Finally, we have a hope of seeing the 'forest' and how we are changing it.

Stuart L. Pimm

1 Introduction

Ecology is the study of the patterns of abundance and distribution of living organisms in relation to their environment. This broad subject area is studied from many points of view by various subdisciplines of ecology. Physiological ecologists attempt to understand the physiological adaptations of individual organisms to their environment and behavioural ecologists study the corresponding behavioural adaptations including social behaviours. Population ecologists are concerned with the causes of change in population numbers. Community ecologists look for patterns in the complex webs of interacting species. Ecosystem ecologists focus on the flows of energy and matter through ecosystems and the biological processes that regulate these flows. Evolutionary ecologists seek to understand how natural selection leads to the patterns of adaptations of biological species and the observed relationships among species.

Ecologists realize that all of these areas of study are deeply interrelated and that the subdisciplines in some sense artificially divide something that should be studied as a whole. However, from both the theoretical and empirical viewpoints, it is difficult to unite in a single study all of these aspects of ecology at once. The difficulties in being cognizant simultaneously of the relationships among such disparate aspects of ecology as the flow of energy and material, population and community dynamics, and behavioural patterns of individuals are not just practical, but conceptual as well. Therefore, despite the basic holistic orientation of ecology, individual ecologists, like scientists in other fields, tend to specialize. They often adopt a point of view in which one or another of the aspects of ecology becomes the central one. The result is, as Robert May stated, 'Ecological theory comes in many forms' (see McIntosh, 1985).

The study of ecological food webs, an important aspect of the biotic community, is an example of an active area of research. This area of theory, sometimes referred to as the 'population-community' approach, focuses on the dynamics of population interactions and the patterns of trophic connections (who eats whom) among species. The heritage of food web theory extends back to Lotka (1956) and Volterra (1931). Expositions of modern research in the theory of food webs can be found in books by May (1973) and Pimm (1982) and in papers such as that of Fretwell (1987), though these reviews hardly begin to exhaust the diversity of work in this area.

A second major area of inquiry, the study of flows of energy and material in ecological systems, or the 'process-functional' approach, has a history of similar length. Lotka (1956) presented detailed descriptions of a number of material cycles in the biosphere, based on earlier work. Lindeman (1942), Hutchinson (1964) and H. T. Odum (1957) gave this approach a renewed impetus which has influenced research to the present. Further reviews of work on fluxes of energy and material in ecological systems include those of Phillipson (1966), Morowitz (1968), Odum and Odum (1976), Likens (1985) and Ulanowicz (1986). Although there are differences in style and purpose of these authors, they seem to agree on the fundamental necessity of understanding energy and material fluxes in order to understand the behaviour of ecosystems.

These two aspects of ecological systems, food web dynamics and energy and material flows, are interrelated. On the one hand, energy and material flow is essential to plant and animal populations, so that the structures of food webs are influenced by limitations of energy and matter, as was recognized by Elton (1927) and Lindeman (1942). On the other hand, interactions among species populations within food webs may themselves influence energy flow and biomass production (e.g., Hairston *et al.*, 1960; Fretwell, 1977; Oksanen *et al.*, 1981; Carpenter *et al.*, 1985; DeAngelis *et al.*, 1989).

Energy is so overwhelmingly important in nature that ecologists have generally emphasized its central role in food web patterns and dynamics. However, an enormous effort has been expended on the study of material elements that are essential to plant and animal growth, called nutrients or bioelements, indicating that they are frequently the factors limiting growth in ecological systems. The elucidation of the mutual influences between the structure and dynamics of food webs and the flow of limiting nutrients through them is an important area of research today. This book attempts to synthesize results of studies on this interface, both by using what can be learned from mathematical models and by organizing the available empirical data around a theoretical framework provided by the models. I start on this task by outlining, in the remainder of this chapter, some elementary ideas about both bioelement cycles and food webs.

1.2 ENERGY AND NUTRIENT FLOWS IN THE BIOSPHERE

One of the basic relationships in nature is that between energy and the flow of matter. The water or hydrological cycle, since it is relatively well known, is a useful example for illustrating this relationship. Water flows downhill, as long as it is not impeded, decreasing its gravitational potential energy. The reason that all of the earth's water does not stay forever in the lowest places on the earth's surface is that solar radiation causes evaporation of surface water and also creates atmospheric convection currents by heating air near the surface. The warm air is less dense than the cooler air above it, and so is buoyant. As the warm air rises, carrying moisture with it, it expands and cools. If the air cools to the dew point, the moisture condenses into clouds, releasing energy in the process. Under the right conditions, the cloud droplets combine into larger droplets or ice crystals, which fall as precipitation to the ground, completing the cycle.

A continual supply of energy from the sun is necessary to maintain the hydrological cycle. If this energy were shut off, the dynamic processes of water evaporation and precipitation, as well as the complex circulations of water in the oceans and atmosphere that are driven by temperature gradients produced by the sun, would slow down and finally stop (aside from possible motions that could result from temperature gradients caused by the high temperatures of the interior of the earth). For this reason, the earth's hydrosphere and atmosphere are an 'open' system with regard to energy. There is an input of energy from the sun and an output, which balances the input on the average, back into space by reflection and reradiation. For example, much of the energy released by the condensation of moisture in clouds is lost directly as long-wavelength (infrared) radiation into space. This output of energy into space is as necessary as the input of energy from the sun for the maintenance of the earth's cycles of material flow in the approximate steady-state conditions we observe. Material cycles are sustained by the continual inputs and losses, or continual 'throughflow', of energy.

The property of 'openness' of the hydrological cycle with respect to energy contrasts with the 'closed' nature of this cycle with respect to the flow of water. The same constituent atoms of hydrogen and oxygen are used over and over again in this cycle. This illustrates the fundamental difference between the coupled flow of energy and material in the earth's bioelement cycles. New inputs of energy are necessary to do the work of maintaining the cycles, but the same atoms stay in the cycles. Energy that has performed work and is reradiated back into space at a longer wavelength (lower frequency) than the input solar radiation is referred to in thermodynamics as 'degraded'. Though the energy is conserved, energy that has performed work is transformed. In technical terms, its thermodynamic 'free energy' decreases, so that it is less able to perform further work. The atoms of the bioelement cycles, however, remain unchanged no matter how often they are recycled.

The vastness of the hydrological cycle is difficult to appreciate fully. Taube (1985) pointed out that energy flow from the sun evaporates about 1.6×10^{10} kg of water per second, requiring solar power of 40.8×10^{15} watts (or 40 800 terawatts). The average water molecule stays in the atmosphere about nine days before returning to the earth as some sort of precipitation or condensation.

The hydrological cycle is only one of many material cycles at the surface–atmosphere interface. All other material cycles are analogous to the hydrological cycle in requiring energy input over parts of the cycle and releasing energy over other parts of the cycle. Many of these material cycles, including those of the bioelements vital to life, are intimately connected with the hydrological cycle, because the material atoms and molecules spend at least part of their cycles in aqueous solution.

The zone of the earth that extends from the bottom of the sea and from the lowest organic layers of soil up to the highest elevations in the atmosphere to which living organisms can be carried by wind is referred to as the 'biosphere'. This band of several tens of kilometres lies within the band occupied by the hydrological cycle. The hydrological cycle is altered in many ways by the presence of life, but would exist on earth whether or not there were living organisms. In the absence of life, however, many other cycles of the material atoms classified as bioelements would either not exist or would be quite different from their present forms. The carbon cycle, for example, despite its having important purely geochemical components (Berner and Lasaga, 1989) is profoundly influenced by photosynthesis in plants (or primary producers, or autotrophs), which creates complex carbon compounds. Carbon dioxide (CO_2), which is a gaseous component of the atmosphere, diffuses into plants during the times of day and seasons of the year when the plants are photosynthesizing. Light quanta (photons) are stored as high-energy compounds by photosynthesis. In this process light photons produce excited electrons and the chemical substance nicotinamide adenine dinucleotide phosphate, NADP, which is an electron acceptor, is converted to the reduced form, $NADPH_2$, which can then reduce CO_2 to carbohydrates.

The carbon cycle is completed when living biomass respires or when dead biomass decomposes, releasing CO_2 back to the atmosphere. Photosynthesis, or the chemical reduction of CO_2, is the 'uphill' side of the carbon cycle, in analogy to the evaporation of water in the hydrological cycle. The oxidation of carbohydrates represents the 'downhill' side of the carbon cycle, analogous to the condensation of water in the hydrological cycle.

In terms of energy, the global carbon cycle is miniscule compared with the global hydrological cycle, consuming an average of 92 terawatts of power globally, compared with the 40 800 terawatts of the hydrological cycle. Correspondingly, the hydrological cycle is much more important to living organisms than the organisms are to the hydrological cycle. The flow of water through terrestrial plants and into the atmosphere as vapour carries about 815 g of minerals for the synthesis of each 3.1 kg of dry biomass (Taube, 1985). This includes nitrogen, phosphorus, potassium, calcium, magnesium, iron, chlorine and the other chemical elements indispensable for life.

On a global scale, many of the bioelement cycles of the biosphere, like the water cycle, are almost closed. This is particularly true for nutrients that form volatile gases, such as carbon (CO_2), nitrogen (N_2, NH_3) and sulphur (H_2S, SO_2). Non-volatile nutrients like phosphorus tend to travel unidirec-

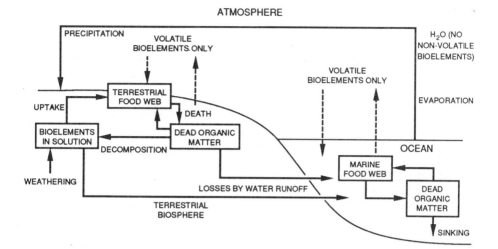

Figure 1.1 General schematic of bioelement cycling on a global scale. Movement of non-volatile elements, such as phosphorus, is largely one way, towards ocean sediments.

tionally from land towards ocean sediments (Figure 1.1), though over geological time scales the uplifting of land may replenish the terrestrial supply of non-volatile bioelements. Even the cycles of volatile bioelements are not strictly closed because these elements are added to the surface in small amounts by volcanic activity (and, increasingly, through fossil fuel burning, especially in the case of carbon and sulphur), while some are continuously being buried in sediments in the deep sea and thus removed from cycling in the biosphere.

A bioelement cycle can in principle be closed because the individual atoms are indestructable and can be used again and again in living organisms. Energy used in the biological processes, in contrast, is 'degraded' during the performance of work, as it is in performing the work of maintaining the hydrological cycle. For example, when a plant is consumed by a herbivore, organic compounds in the plant biomass are broken down and the energy and material released are used to synthesize herbivore biomass. Much of the energy is released as thermal radiation, or heat, which ultimately radiates into space. In general, less than about 10% of the energy embodied in plants gets transformed into herbivore biomass, and the energy that is lost can never be used again. It is true that some energy from dead herbivores and carnivores recirculates through the detrital food chain, but this amount must, by the laws of thermodynamics, be relatively tiny compared with that coming directly from primary production. In contrast, nutrient atoms continue in circulation and can be reused by autotrophs, herbivores and carnivores an indefinite number of times.

Because of the fundamental difference between energy and bioelements, these flows are represented differently in a general diagram of processes in the biosphere (Figure 1.2). In this figure, which is a variation on one presented by Morowitz (1968, Figure 4–1), solid lines show the cycling of material from abiotic pools through living matter and back again, while dashed lines indicate the influx of solar radiation, the transfer of chemical energy and the efflux of thermal radiation to outer space which drives the system.

A characterization of the earth's global material cycles is an important goal of geologists and ecologists. This is especially true since human activities are altering crucial cycles such as the carbon cycle, with possible consequences for global climate. These global cycles, however, represent a summation over processes occurring in the great diversity of local ecosystems, each with its own pattern of energy and material fluxes and each with its own unique biotic community. Chapter 12 of this book will return to

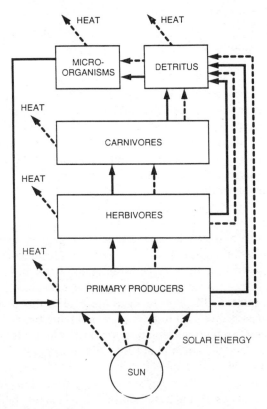

Figure 1.2   General representation of energy flows (dashed lines) and material cycles (solid lines) in the biosphere. The energy flows included are solar radiation, chemical energy transfers (in the ecological food web) and radiation of heat into space. (Adapted from Morowitz, 1968.)

Table 1.1 Nitrogen inputs and losses from different forest ecosystems. (From O'Neill and Reichle, 1980)

System	Country	Inputs[a] (kg ha^{-1} year^{-1})	Losses[a] (kg ha^{-1} year^{-1})
Temperate deciduous[b]	USA	13	109.5
Temperate deciduous[c]	Belgium	8.7	79
Tropical rain[d]	Puerto Rico	49	160
Coniferous[e]	Sweden	8	58
Tropical dry[f]	India	51	133
Mediterranean evergreen oak[g]	France	1.47	72
Montane coniferous[h]	USA	5.1	12.3

[a] Inputs include atmospheric inputs and nitrification minus denitrification. Losses indicate total nitrogen lost from vegetation through litterfall, mortality, consumption and root death.
[b] Henderson and Harris (1975).
[c] Denaeyer and Duvigneaud in Reichle *et al.* (1973).
[d] Odum and Pigeon (1970).
[e] Anderson in Reichle *et al.* (1973).
[f] Bandhu in Reichle *et al.* (1973).
[g] Lossaint in Reichle *et al.* (1973).
[h] Stark (1973).

global cycles, but the intervening chapters will focus on the smaller scale relatively homogeneous ecosystems that ecologists have measured and analysed.

As an example of the fluxes associated with particular ecosystems, consider the measured yearly nitrogen inputs and outputs per hectare for seven different forest ecosystems shown in Table 1.1. The inputs listed include atmospheric inputs and nitrification minus denitrification. The losses are the sum of litterfall, tree mortality, consumption and root death and so do not represent nitrogen lost from the system as a whole. In small-scale ecosystems, the relative magnitudes of nutrient inputs from and outputs to the external environment are crucial. The losses of nitrogen from living biomass shown in Table 1.1, since they are so much larger in all cases than the input of new nitrogen to the system from atmospheric input and fixation, can only be sustained if much of the biomass production is sustained by recycling of nitrogen.

1.3 NUTRIENT CYCLING IN ECOSYSTEMS

The ecosystem has been defined as

'any unit that includes all of the organisms (i.e. the community) in a given area interacting with the physical environment so that a flow of energy leads to clearly defined trophic structure, biotic diversity and material cycles (i.e. exchange of materials between living and non-living parts) within the system' (Odum, 1971).

This definition turns out to be rather imprecise when one is attempting to develop a rigorous theory of ecosystems. There are no actual 'units' of space that are as distinctly bounded as the units of study of other biological fields – the individual organism or the individual cell, for example. Even for such an apparently discrete entity as a lake, for example, it is not obvious exactly where one should choose the boundaries. The lake is closely connected to its watershed through inputs of inorganic and organic materials and may be linked to other lakes through overland and groundwater flows. The problem is even more difficult in deciding how much of, say, a forest, or how much of the open ocean must be considered for the term 'ecosystem' to apply. Questions of scale are, therefore, important issues in ecology. A few hectares of area might suffice as a 'unit ecosystem' if all of the organisms are relatively small and sedentary, but not if it is desired to include far-ranging animals or sparsely scattered plants.

These problems of precise definition will not concern us here. I will take the practical view that a biological system can be considered as an ecosystem if it is self-sustaining within certain limits; in particular, if it is of sufficient scale to contain both organisms that photosynthesize light to produce biomass (autotrophs), and decomposing organisms that transform the biomass back into abiotic nutrients available for use by plants again. The sunlight needed to power the photosynthesis must, by definition, come from outside the ecosystem and I will assume that there can also be inputs and outputs of abiotic elements, including water and nutrients. These specifications are broad enough to allow systems as small as laboratory microcosms (e.g. Whittaker, 1961; Maguire *et al.*, 1980) to be called ecosystems.

Ecologists who study the flows of energy and matter while largely ignoring the population dynamics of individual species follow the process-functional approach in studying ecosystems and usually choose carbon or energy as fundamental units rather than numbers of organisms, and may lump individual species into functional groups. This sort of approach can be seen in Lindeman (1942), Morowitz (1968), Van Dyne (1969a,b), Patten (1971), Odum (1971), Wesley (1974), Ulanowicz (1986) and others. Ecologists interested in questions in which population numbers play a significant role either in the interactions (e.g. host–parasite systems, lynx–hare cycles) follow the population-community approach and usually use population numbers as variables. This approach is very prevalent in current texts and monographs (e.g. MacArthur and Connell, 1966; May, 1973, 1976; Maynard Smith, 1975; Hassell, 1978; Begon and Mortimer, 1981).

The main structural components and processes of a generalized ecosystem are shown in Figure 1.3. An autotroph component creates biomass through

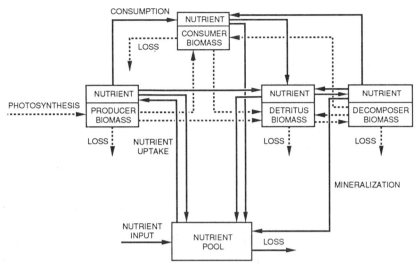

Figure 1.3 General representation of energy and nutrient flows in an ecosystem. Dashed lines represent energy flows and solid lines represent nutrient cycles and inputs and outputs. Outputs of nutrients include losses by water transport of dissolved nutrients and drift or migration of living organisms.

the process of primary production, which involves photosynthesis, the transpiration of water and uptake of nutrient ions and molecules. The consumer or heterotroph component of the ecosystem includes herbivores, carnivores, omnivores, and parasites that consume the biomass of other organisms for use as energy and material for their own biomass. Processes of respiration, death and decomposition by the decomposer component release energy, some of which may be recycled. The decomposers remineralize nutrients back to inorganic forms that can be used again, but can also take up available nutrients, decreasing their availability to autotrophs. Pools of nutrients in the detritus component, soil solution and in solid form are also crucial components of the ecosystem. Movement of nutrients by water, wind and organism transport within the ecosystem as well as inputs to and losses from the ecosystem are also essential processes.

When I use the term ecosystem here, I have in mind this relatively generic concept, though I will also frequently refer to more specific ecosystems, such as lakes, areas of tropical forest or coral reefs. All of the components and processes depicted in Figure 1.3 occur in these specific ecosystems, but the sizes of the components (absolute and relative) and the rates of the processes will differ among specific systems.

By 'nutrients' I mean those elements, the bioelements, that are essential to the structure and/or function of organisms. Although much of the model analysis described here applies to a generic limiting nutrient, the examples usually involve specific nutrients, such as nitrogen, phosphorus, potassium and calcium. These nutrients tend to be ones that are limiting, an aspect that has special implications for system dynamics.

The study of the flows of biologically essential elements between the biotic and abiotic components of a system is referred to as biogeochemistry. Hutchinson (1948), who helped call attention to the biogeochemical cycles in ecology, described the phosphorus cycle in a lake. Phosphorus, an essential constituent of all organisms, usually first enters biological systems by the weathering of rocks. The amount released per unit time from the lithosphere is only a small fraction of what is needed in biological production, however. Most of that used in production is recycled or regenerated phosphorus, available to be taken up again after organic matter is broken down into phosphates. Eventually, phosphate that has been part of the biosphere finds its way to the sea, where it may be buried in deep sediment, so that it is no longer available to life processes.

While phosphorus and many other nutrients come from rocks, nitrogen is stored in huge quantities in the atmosphere (and dissolved in ocean water) as the inorganic compound N_2. However, N_2 is not directly available to plants and it must be converted first to available forms by lightning (nitrates) or nitrogen-fixing bacteria (amides) before it can be used in protein synthesis. Denitrifying bacteria cause nitrates to be lost again to the air by conversion to N_2.

As mentioned earlier, volatile nutrients like carbon (as CO_2), nitrogen (as N_2, NH_3) and sulphur (as H_2S, SO_2) will flow freely between the system and the atmosphere. Other, non-volatile elements, like calcium, phosphorus and potassium, can be carried by water or wind into and out of the system, in addition to being weathered from rocks.

There are four important aspects of nutrient cycling to which special notice should be given at this point, as these have important consequences for the structure and dynamics of food webs. These are (1) nutrient limitation of primary and secondary production, (2) the recycling of nutrients and mechanisms of recycling, (3) chemical complexity and (4) stoichiometry.

Nutrient limitation of autotrophs, as well as of heterotrophs and decomposers, is common in all ecosystems, as will be documented later. This limitation may be compensated for to some extent by internal recycling of nutrients. Depending on the ecosystem, the imports and exports of material may be very large or very small compared with the recycling of material within the system. This important property of the relative degree of recycling versus new inputs of nutrients in ecosystems will be discussed in detail in many subsequent chapters, as it has important effects on the dynamics of the system.

Every nutrient cycle differs from others in its particulars, and some, like the nitrogen cycle, are enormously complex, involving several different forms of compounds. As an example, Figure 1.4 shows some of the chemical transformations occurring within the nitrogen cycle of an estuary (Sprent,

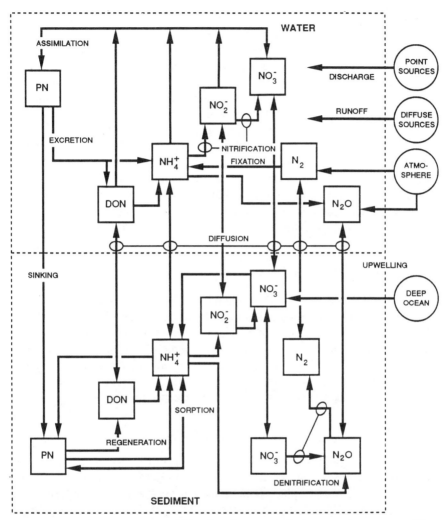

Figure 1.4 Transformations within the nitrogen cycle in estuarine systems. The main processes represented here are assimilation of some forms of inorganic nitrogen into biomass (or particulate nitrogen PN), regeneration of dissolved organic nitrogen (DON) and ammonium (NH_4^+) through excretion and decomposition, fixation of NH_4^+ from N_2 and denitrification of NO_3^-, and conversion between different nitrogen oxidation states. (Adapted from Day *et al.* 1989.)

1987, for more details on the nitrogen cycle). As far as possible, however, I will ignore much of this complexity and simplify a particular limiting nutrient to having a form that is available to uptake by autotrophs and bacteria and a biotic form as part of a plant or animal or detritus. Also, there will always be the possibility of exchanges of the nutrients between the ecosystem under consideration and its external surroundings.

Chemical stoichiometry refers to the proportions of various chemical elements in substances. The proportions of nutrients in biomass display considerable regularity within related groups of organisms (Reiners, 1986). For example, 'protoplasmic life', defined by Reiners to include all cells, from single-celled to multicellular organisms (excluding mechanical structures such as cell walls, shells, skeletons), contains C, H, O, N, P and S in highly regular proportions. The facts that these nutrients occur in fairly strict proportions and form the bulk of protoplasmic life usually means that one or more of these elements will be kept in short enough supply in the pool of available nutrients in an ecosystem to be limiting. Reiners argues that the stoichiometry of biomass thus has a great effect not only on species strategies for sequestering and recycling certain nutrients, but also on the biogeochemical cycles of the earth. The effects of stoichiometry on food web dynamics will also be evident in the examples presented here.

1.4 FOOD WEBS

Shelford (1913) and Elton (1927) were responsible for putting such concepts as 'food chain' and 'pyramid of numbers' firmly into the ecological literature, though these ideas were recognized earlier (see McIntosh, 1985). The term food chain formalizes the observation that some of the energy and matter fixed by photosynthesis in autotrophs is consumed and converted into tissue by herbivores, that some of this biomass is, in turn, consumed by carnivores, and so on. A food chain is some particular sequence of trophic relationships, such as

pine trees → aphids → spiders → titmice → hawks

(Colinvaux, 1986). However, because most consumer species feed on more than one prey species and most prey have more than one species that feeds on it, the network of trophic interactions that occurs in most ecosystems is more appropriately termed a 'food web'. The food web characterization of an ecosystem differs from the functional characterization (Figure 1.3) in emphasizing taxonomic types rather than functional components. However, the taxonomic diversity in most ecosystems is far beyond the ability of ecologists to investigate fully. Thus all published food webs are necessarily incomplete. In most cases, species are lumped together into higher tax-onomic entities, such as the food web diagram of Gatun Lake shown in Figure 1.5, in which taxa of a variety of levels from the species to the phylum are put in the same diagram. Even with simplifications such as these, food webs are complex enough to be inaccessible to thorough understanding. Figure 1.5 is shown as an example of a degree of complexity that will not generally be studied here (although some of the examples in Chapter 9 approach such situations). Most of the analysis presented here is restricted to systems of the form of Figure 1.3.

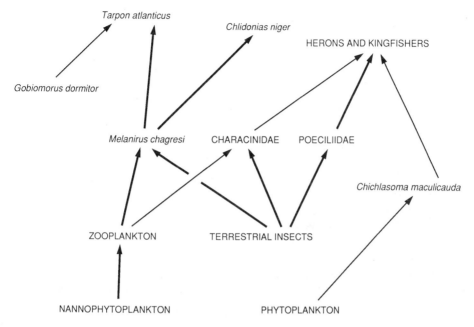

Figure 1.5   A simplified food web hypothesized for Gatun Lake showing the diverse paths that energy and matter can follow (bold lines indicate the more important links). (Adapted from Zaret and Paine, 1972.)

The term 'pyramid of numbers' in food chains refers to the fact that large numbers of small prey are usually necessary to sustain one large predator. This concept was given a more general form, 'pyramid of energy', (or more properly, a pyramid of energy flows) by Lindeman (1942) and others. This reflects the consequences of the Second Law of Thermodynamics that only a small part (usually no more than about 15%) of the energy contained in a resource organism (plant or animal) can be transformed into consumer biomass. The rest of the energy is dissipated as heat energy. The implications of this inevitable loss were grasped by Hutchinson (1959) and others, who hypothesized that progressive degradation of energy up the food chain is the primary factor that limits food chain length from the autotroph to the top carnivore. There are typically only three or four levels observed in such food chains.

Another hypothesis concerning food chain length (or, more generally, the shortest chain from the autotroph to the top carnivore in a food web) is that it and other structural aspects of these systems are governed not by energy dissipation but by dynamic characteristics of the interacting populations in the web. Pimm and Lawton (1977, 1978) and Pimm (1979, 1980a,b, 1982) have been supporters of this view. These two alternative hypotheses will be discussed in more detail in Chapter 8.

It may also be true that some factors whose importance in food web structure has not yet considered may play a role. The effects of the limitation of one or more key nutrients may be as important as other factors in influencing ecosystem structure (including food chain length) and stability of many ecosystems. Moreover, the ecosystem structures that have evolved in some ecosystems to conserve and recycle nutrients may have their own effects on system stability.

1.5 USE OF MATHEMATICAL MODELS

Like other scientists, ecologists investigate their subject by the use of models. The models may be purely conceptual, such as the physicist's idealization of an ideal gas being composed of tiny hard elastic balls, or mathematical, as when equations for the kinetic energies of the atoms of a gas are related to measurable properties of the gas.

Whether based on population numbers or biomass or energy, all the ecological models mentioned earlier can be classed into one of three general types, labelled 'descriptive' models, 'explanatory' or 'heuristic' models and 'predictive' models. These types may be defined as follows:

Descriptive models

These models attempt only to describe a set of observations in mathematical form, for example, by fitting a curve to a set of points. Most allometric relationships in biology are of this type. They simply codify empirical observations that some measurable quantity, say rate of respiration, when plotted against another relevant quantity, say animal size, on a log–log plot, forms a relatively straight line. No explanatory mechanism is built into this model, although the model itself may be used to suggest possible mechanisms.

Explanatory models

These attempt to explain observed data in terms of more basic known mechanisms. The Lotka–Volterra predator–prey population model attempts to explain observed oscillatory behaviour of certain species by proposing that the rate of consumption of prey by the predator population is proportional to the product of the sizes of the two populations. The emphasis of the model is on showing that the oscillations can exist, rather than on predicting precisely the values of the population levels at a given time.

Predictive models

The third major type of model is the predictive model. Models of this type tend to be very complex. A prediction of tomorrow's weather, for example, may be made from a general circulation model of the atmosphere, with thousands of variables. Similarly, in ecological systems complex simulation models with many compartments are often used to attempt to make precise predictions.

These three functions of description, explanation and prediction are usually not clear-cut and most modelling efforts have some aspects of all three. The present book is aimed at providing heuristic explanations. Thus models will be used to help understand the interaction of nutrient cycling and food web dynamics.

1.6 SCOPE AND STRATEGY

The interactions between material cycling and food web dynamics are highly complex. In fact, it will be shown that even quite simple model systems can display behaviours that are not easily anticipated. This is because the relationship between nutrient flow and the biomasses of functional groups in food webs is one of mutual causality. Biomass levels and rates of change are affected by the availability of limiting nutrients, but the sizes of the biomasses and overall food web structure in turn affect the flow and hence availability of nutrients. For this reason, any attempt to understand such systems should ideally start at a basic level and work up to great complexity gradually. That is the strategy of this book.

Chapter 2 discusses basic concepts of flow of a bioelement in solution into and out of a single compartment, say an idealized homogeneous lake. This is a simple enough system to treat analytically and to allow clear definitions of a number of basic concepts such as input loading, dilution rate, steady state, transient behaviour, return time to steady state, perturbation, resistance, local stability and resilience. This is extended to a two-compartment model, where the added compartment acts like the sediment layer of a lake or the ion exchange clay in soil. This system can still be studied analytically, but it exhibits some new behaviours.

Chapters 3 and 4 consider the case where there is a single autotrophic species and a litter or detritus compartment in addition to a pool of available nutrient. In the first of these two chapters the growth rate of the autotroph is assumed to be limited by the nutrient concentration in the available pool. In Chapter 4, a more complex representation of autotroph growth is examined, the cell quota model, in which the growth rate is limited by a variable internal nutrient concentration. Properties of steady-state solutions, local stability and resilience, as well as the possibility of behaviours such as catastrophies, are examined.

In Chapters 5 and 6 herbivores are added to the system and a variety of effects of herbivore–autotroph interactions are considered in combination with nutrient limitation and recycling. The possibility of top-down control of the autotroph by the herbivore introduces new dynamic behaviours in such systems. Other new phenomena arising from the added complexity include the effects of food quality for the herbivore (nutrient concentration in food) on system stability.

The effects of detritus and the detritus-based food chain, ignored to this point, are considered in Chapter 7. New modes of behaviour arise from such new complications as competition for nutrients between autotrophs and decomposers for nutrients. Chapter 8 generalizes the results of earlier chapters to large food chains and webs, establishing some general principles of these systems. Competition for nutrients by different autotroph species and the effect of higher trophic levels on this competition are added in Chapter 9.

The first nine chapters together follow the plan of gradually increasing complexity up to the point of a full-scale food web. These chapters do not begin to exhaust interesting aspects of the subject of nutrient flow and ecological dynamics, however. Chapter 10 considers the effects of disturbance and fluctuations in nutrient availability and how these affect food web dynamics and, in particular, the coexistence of competing species. Chapter 11 focuses on a special aspect of food webs in relation to nutrient fluxes that is of great importance, spatial extent and heterogeneity. Chapter 12 reviews some of the applications of nutrient cycling and food web theory to environmental problems. It also examines a current environmental problem of particular importance, possible change in global climate, as an example of the vital importance of understanding the interrelationships of nutrient cycling and food web dynamics.

2 General concepts of nutrient flux and stability

2.1 INTRODUCTION

Water can carry a wide array of dissolved substances, including every bioelement [human urine, for example, contains up to 80 dissolved compounds (Botkin, 1990)]. Thus the fluxes of many important nutrients are closely connected with the hydrological cycle, nutrients with gaseous forms such as N_2 and CO_2 being partial exceptions. The concentration of any given dissolved substance in the waters on the earth's surface is variable, both in space and time. The dissolved substances in a particular unit volume of water depend on the recent history of that water. Passage of water through rocks and soil dissolves chemical compounds, whole processes such as diffusion, the mixing of water from different sources, and evaporation causes changes in concentrations. These processes go on even in the absence of biota, and it will be useful to begin the mathematical description of these nutrient dynamics in physical systems without the complication of biota.

The mathematical approach used through much of this book is that of compartmental modelling. This approach has been described in detail by Sheppard (1962), Jacquez (1972) and Anderson (1983), among others. It assumes that a system can be divided into a finite number of compartments or pools, each of which is internally homogeneous and connected to other compartments. Associated with each compartment is a variable that represents an amount of matter, energy or nutrients in that compartment at a given time. A differential equation describes the dynamic changes in the amount of matter, energy or nutrient in each compartment through time, as these are affected by fluxes into and out of the compartments. As an example of a compartment, think of a well-mixed lake containing uniformly distributed dissolved phosphorus and no biota. Streams flowing into the lake bring input fluxes of phosphorus and outflows carry out phosphorus. The changes in the amount of phosphorus in the lake depend on the instantaneous difference between inflows and outflows. This is an idealization, since no body of water containing perfectly homogeneously distributed dissolved substance exists in reality. However, it is a useful idealization that may often be accurate enough to serve as a good first approximation. It is necessary, in any case, to understand the dynamics of such imaginary systems before one can go on to more realistic situations.

Although the most common application of compartmental models (often called 'box and arrow diagrams') has been to quantitative areas of physiology, such as pharmacokinetics, compartmental modelling became an important tool in ecology over two decades ago. Perhaps the earliest pioneer in compartmental modelling in ecology was A. J. Lotka, whose book, *Elements of Physical Biology*, published in 1924, contained conceptual models of the nitrogen, phosphorus and other material cycles in the form of a number of connected compartments (plants, animals, carcasses, manure, soil and so forth).

In more recent times compartmental modelling became popular under the name of 'systems ecology', developed by a number of ecologists, including E. P. and H. T. Odum (1955), who modelled the energy flow in a coral reef community, J. S. Olson (1965), who modelled flows of caesium in a forest, and G. Van Dyne (1966), who developed early models of energy flow in grasslands. Since that time, it has been customary to use compartment models to describe and predict the movement of substances through ecosystems. Here we will start with the simplest compartmental model, one with a single compartment.

2.2 SINGLE COMPARTMENT OR POOL

In modelling the flow of matter, one starts with basic conservation principles. The principle of conservation of mass states 'the rate of change of mass in a specified region of space equals the rate at which mass enters that region minus the rate at which the mass leaves' (Denn, 1986).

Let us apply this to a situation in which there is a constant volume V (for example, 1 cc) of water, containing solute of initial concentration C_0, in a pool (Figure 2.1). Let water, at a constant rate, $q(l\,s^{-1}$, where 1 litre, l, is

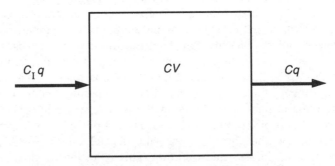

Figure 2.1 Single-compartment model of a solute of concentration C in a water body with an inflow $C_1 V$ and an outflow CV of solute. If the initial concentration is C_0, the time course of concentration is given by Equation (2.3).

approximately 1 kg in weight), enter the pool and let there also be the same constant outflow q. Assume that at time $t = 0$ a dissolved solute with concentration C_1 (g solute l^{-1}) is present in the input water and assume further that the pool is rapidly mixed. The solute is assumed to be low enough in concentration to make up a negligible fraction of the volume in the pool, input water and outflow water. The problem now is to determine the concentration of the solute in the tank, which we will define as $C(t)$, for all times $t > 0$.

A straightforward way to approach this problem is to consider the water in the pool to be a compartment and to apply the mass conservation law to the total mass of solute in the compartment. This total mass is the concentration $C(t)$ times the volume of water in the pool V, or CV (I will usually abbreviate $C(t)$ to C from now on). The inflow of mass of solute, or the *input loading* is $C_1 q$ and the outflow is Cq. The law of conservation of mass then states

(Rate of change of mass of solute contained in compartment)
    = (rate at which mass enters) − (rate at which mass leaves)

In symbolic terms,

$$d(CV)/dt = C_1 q - Cq \tag{2.1}$$

where $d(CV)/dt$ represents the rate of change of CV. The quantities V, q and C_1 are all constants; thus, $dV/dt = 0$, so $d(CV)/dt = VdC/dt + CdV/dt = VdC/dt$, and Equation (2.1) becomes

$$VdC/dt = q(C_1 - C) \tag{2.2}$$

Equation (2.2) is a linear equation because in the terms where C occurs it is always as the first power of C. The equation can be solved to obtain the solution

$$C = C_1(1 - e^{-(q/V)t}) + C_0 e^{-(q/V)t} \tag{2.3}$$

which can be checked by substitution of C from Equation (2.3) into Equation (2.2). In Equation (2.3) C_0, formally a constant of integration, represents the initial condition on concentration at time $t = 0$. The solution (2.3) is a general solution of the differential equation in the sense that C_0 can be chosen to fit all possible initial conditions of C. The constant C_1 is the steady-state equilibrium concentration, to be discussed below. The ratio q/V, the exponential coefficient of time, t, sets the time scale of the dynamics of the variable C through time. This coefficient of time in the exponent is of special importance in linear differential equations and is referred to as an **eigenvalue**. A first-order linear differential equation such as Equation (2.1) has one eigenvalue. Eigenvalues will be encountered frequently throughout this text. They become especially important in describing the dynamics of multicompartment systems, which have as many eigenvalues as compartments. Concentration C given by Equation (2.3) is plotted in Figure 2.2 for

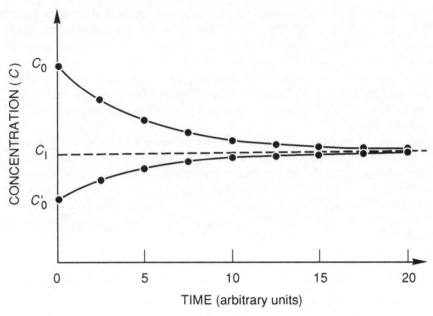

Figure 2.2   Plot of concentration C as a function of time for two different starting concentrations, C_0 and C_0' [from Equation (2.3)]. The ratio q/V is set to 0.20. $T_R = 5.0$ is the return time to a fraction e^{-1} of the initial deviations $(C_0 - C_I)$ and $(C_0' - C_I)$

both the case in which $C_0 > C_I$, meaning that the concentration is started out above the steady-state level, and that in which $C_0' < C_I$. In both cases the solution approaches C_I through time.

2.3   PROPERTIES OF THE SOLUTION

The solution, Equation (2.3), though very simple, contains many of the basic properties that we will be interested in trying to look for in more complex models. These are as follows.

Steady-state equilibrium

As $t \to \infty$, the concentration approaches a constant value

$$C^* = C_I \qquad (2.4)$$

where the * signifies that the variable is at steady-state equilibrium.

Transient behaviour

The part of the solution (2.3) that depends on t is called the transient component of the solution:

$$(C_0 - C_1)\,e^{-(q/V)t}$$

This transient component damps away or 'decays' as time t increases, becoming quite small as t reaches large values (several multiples of V/q), leaving only the steady-state component of the solution, C_1.

Stability

For any starting value of C, C_0 (C_0 can plausibly only be zero or a positive value), the solution returns to $C^* = C_1$ (Figure 2.2). Thus, the system is both locally and globally stable. **Local asymptotic stability** (also called Lyapunov stability) means that there is at least a small region around C^* such that, following a perturbation of C to any point in this region, the solution returns towards C^* ('asymptotically' meaning that the solution literally never reaches C^*, because $e^{-(q/V)t}$ is always positive, no matter how large t is). **Global asymptotic stability** means that the solution returns asymptotically towards C^* for all possible starting values of C. In linear systems such as Equation (2.1), local stability always implies global stability. Many systems in nature, however, are non-linear and some of these may be locally stable but not globally stable. Such systems may not return to the same equilibrium following a large perturbation. Cases of this will occur in some of the complex biotic systems studied later.

Resilience

Resilience is a type of stability, sometimes called 'relative stability'. Assume that a system is at least locally stable and so will return to steady-state equilibrium following a perturbation. Resilience is a measure of the rate at which the system approaches steady state following the perturbation. To formalize this somewhat, suppose as before that the concentration in Equation (2.1) is perturbed to C_0. How quickly does the perturbation decay to some fraction of its original value? (It is conventional to take as this fraction e^{-1}, where $e = 2.718\ldots$ is the natural logarithm.) From the solution, Equation (2.3), it is clear that when $e^{-(q/V)t}$ becomes equal to e^{-1}, the initial deviation $(C_0 - C^*)$ will have returned to a fraction e^{-1} of C^*. This occurs at a time $t = T_R$, where

$$T_R = 1/(q/V) = V/q \qquad (2.5)$$

The value T_R is called the **return time to equilibrium** and its inverse is a measure of resilience:

$$\text{Resilience} \sim 1/T_R \qquad (2.6)$$

An equivalent way to calculate the return time to equilibrium, T_R, which will be useful in much more general applications, is to integrate the area under the curve in Figure 2.2, normalized by the initial size of the perturbation away from C^*, $C_0 - C^*$. The appropriate integral for the one-compartment system under consideration is

$$T_R = \int_0^\infty dt(C - C^*)/(C_0 - C^*)$$

$$= [(C_0 - C^*)/(C_0 - C^*)] \int_0^\infty dt\, e^{-(q/V)t}$$

$$= -(V/q)\, e^{-(q/V)t}\big|_0^\infty = V/q$$

which is exactly the same as Equation (2.5). This expression is generalizable to a system of n compartments, with concentrations $C_i(t)$, where i is the compartment number, by performing summations within the integral

$$T_R = \frac{1}{\displaystyle\sum_{i=1}^{n}(C_{i,0} - C_i^*)^2} \int_0^\infty dt \sum_{i=1}^{n}(C_i(t) - C_{i,0})^2 \tag{2.7}$$

Expression (2.7) can also be used for non-linear systems, in which the $C_i(t)$ values cannot be solved for analytically, but can be determined numerically by means of computer simulations.

Residence time

Assume that a molecule of solute has just entered the compartment pictured in Figure 2.1. What is the expected time it will reside in the system given that the compartment is in steady state? Since C^*V is the amount of solute mass in the compartment, and hence is a measure of the number of solute molecules, and C^*q is a measure of the number of solute molecules both entering and leaving the compartment per unit time, the average time any one molecule stays in the compartment is

$$T_{res} = (C^*V)/(C^*q) = V/q \tag{2.8}$$

where T_{res} is called the **residence time**. The mean residence time, also called the turnover time of the system, is equal in this case to the return time, T_R.

The mean residence time can be thought of as a measure of the rate of flushing or dilution of the solute. If there is an initial concentration, C_0, in the compartment and the input water starting at time $t = 0$ is pure (without solute), then the concentration at any later time is described by the equation

$$dC/dt = -(q/V)C \tag{2.9}$$

the solution of which is

$$C = C_0 e^{-(q/V)t} \tag{2.10}$$

At time $t = T_R$, $C(t) = C_0 e^{-1} \cong 0.34 C_0$, at time $t = 2T_R$, $C(t) \cong 0.12 C_0$, and so forth.

2.4 EXAMPLE OF RESIDENCE TIME CALCULATIONS

Canfield *et al.* (1984) computed the mean residence times for several bio-elements in Acton Lake in Ohio (Figure 2.3). This calculation is not as simple as it would appear because the stream inputs are variable both in flow rates and concentrations. Thus one cannot use an instantaneous value of input, such as q in Equation (2.8).

The volume of the lake was estimated to be $95.5 \times 10^5 \, \mathrm{m^3}$ and this volume was assumed to be well mixed, except for some vertical stratification. The water residence time was calculated by summing daily mean stream discharges into the lake until a volume of water equal to two whole lake volumes had been discharged into the lake (water loss due to evaporation was assumed negligible). The average water residence time was assumed to be half of this measured time.

Stream solute inputs were obtained by calculating flow-weighted concentration averages for the streams. An average input concentration (mean concentration in $\mathrm{mg \, l^{-1}}$), C_I, was obtained for each bioelement, and the stream water discharge rate, $q(\mathrm{l \, d^{-1}}$ where d stands for days), was calculated. The total mass (mg), N_T, of a particular bioelement in the lake was found by sampling three vertical layers of the lake. The average concentration of each layer j, C_j, was multiplied by the layer volume, V_j, and then the masses of the layers were summed, or

$$N_T = \sum_{j=1}^{3} C_j^* V_j \, (\mathrm{mg}) \tag{2.11}$$

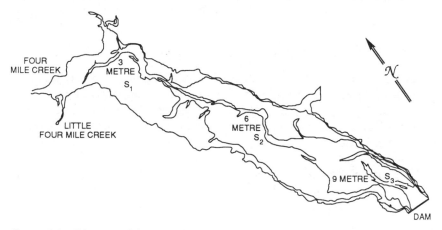

Figure 2.3 Diagram of Acton Lake, Ohio.

Table 2.1   Residence times (days) for selected elements in Acton Lake (from Canfield *et al.*, 1984)

Ca	Mg	Cl	Na	K	Fe	Mn
77	104	105	105	144	930	1800

The residence time, in units of d is, in analogy with Equation (2.8),

$$T_{res} = N_T/(C_1 q)\,(\text{d}) \tag{2.12}$$

The results of the residence times for one of four of the dates for which these were computed by Canfield *et al.* (1984) are shown in Table 2.1.

Note that calcium (Ca) has the shortest residence time, while iron (Fe), manganese (Mn) and potassium (K) are the longest, and magnesium (Mg), chlorine (Cl) and sodium (Na) have intermediate values. These values indicate the degree to which the bioelements react within the lake. Na and Cl are known to be fairly non-reactive in the lake and hence demonstrate 'conservative' behaviour, that is, behaviour expected in terms of the conservation equation (2.1). Mg also seems to exhibit conservative dynamics. Some of the elements displaying long residence times, such as Fe and Mn, may have already been stored in large amounts in bottom sediments and have been undergoing resuspension from these sediments during the period of measurement, thus indicating a relatively high concentration in the water column of the lake at an earlier time. Ca, which showed a relatively short residence time, may have been precipitating during the time of measurement as calcite, so that the water column concentration was correspondingly low.

The study by Canfield *et al.* (1984), and others like it, show that it is insufficient simply to model only the compartment and its hydrological inflows and outflows even in the idealized abiotic case. The interactions of the solute with the sediment must also be considered. An extension of the model to do this is considered in the next section.

2.5   LOSS RATE TO SEDIMENTS

Nutrient retention is the percentage of nutrient supplied to a lake that does not leave via outflow. It can be studied by means of a slightly more complex equation that considers nutrient exchange with the sediments. One particular nutrient budget model is that introduced by Vollenweider (1969), with explicit reference to phosphorus. Here I will follow Chapra's (1975) presentation based on this model, using my own notation.

Suppose that, in addition to stream inflows and outflows, there is a loss to the sediments. This can be represented by another term, subtracted from the right hand side of Equation (2.1), so that

$$V\,\text{d}C/\text{d}t = I_n - qC - vAC$$

or

$$dC/dt = (I_n/V) - [(q + vA)C]/V \qquad (2.13)$$

where

A = surface area of the lake (m^2), assumed to be the surface area of the sediments as well, v = the apparent settling rate of total phosphorus ($m\ d^{-1}$), and I_n = shorthand for the input of nutrient mass per unit time, or qC_1 ($mg\ m^{-2}\ d^{-1}$).

The solution of Equation (2.13) is

$$C = [I_n/(q + vA)]\{1 - e^{-[(q+vA)/V]t}\} + C_0 e^{-[(q+vA)/V]t} \qquad (2.14)$$

This solution, which is plotted in Figure 2.4 for zero and non-zero values of loss rate to the sediments, vA, is different in two important ways from Equation (2.3), in which the loss rate to the sediments was ignored (see Figure 2.4). The steady-state value ($t \to \infty$) is now

$$C^* = I_n/(q + vA) \qquad (2.15)$$

reflecting the fact that the loss rate to the sediments, vAC, causes the

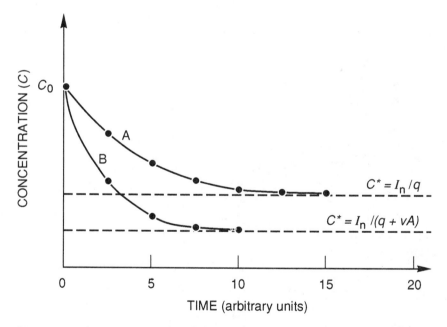

Figure 2.4 Effects of loss rate to the sediments, vA, on the nutrient concentration as a function of time. In curve A, the loss rate to the sediments is zero. In curve B this loss rate was chosen to be of the same magnitude as the loss rate q from the system. Note that the loss rate to the sediments affects both the steady-state concentration and the rate of return to steady state.

steady-state value of nutrient in water to be lowered. The second difference is that the return time to steady state, T_R, is now

$$T_R = V/(q + vA) \qquad (2.16)$$

that is, the return time is lowered by a factor $q/(q + vA)$, indicating that resilience has increased. This results from the retentive power of the sediments. When there is no loss rate to sediments, $v = 0$, so that C^* and T_R reduce to the earlier results of Equations (2.4) and (2.5).

Example: From Equation (2.16), it is possible to derive the phosphorus retention coefficient, R_p, of the lake, defined as the rate of mass lost to the sediments, vAC^*, divided by the rate of mass input to the lake:

$$R_p = vAC^*/I_n = v/(q_s + v) \qquad (2.17)$$

where

$q_s \, (= q/A) =$ the rate of water inflow per unit lake area, or areal water load.

Chapra (1975) fitted Equation (2.17) to data on R_p and q_s for 15 southern Ontario lakes (from Kirchner and Dillon, 1975) using least squares. The fit is shown in Figure 2.5, where the resulting settling velocity, v, is 16 m/year.

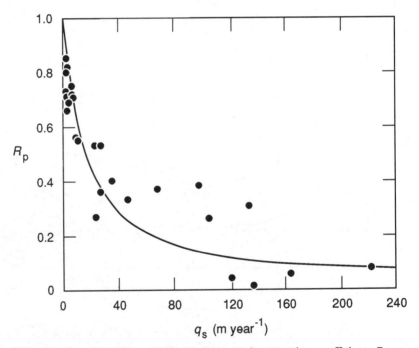

Figure 2.5 Fit of Equation (2.17) to data on the retention coefficient, R_p, as a function of the rate of water inflow per unit of lake area, q_s, from data from 15 southern Ontario lakes. [Adapted from Chapra (1975) based on data from Kirchner and Dillon (1975).]

2.6 FEEDBACK FROM SEDIMENTS

It appears from a number of studies (e.g. Hayes *et al.*, 1952; Rigler, 1978) that there is a net retention of some nutrients, such as phosphorus, in the sediments of lakes, as described in the preceding model. However, not all of the nutrients taken up by the sediments remain there. Some fraction may be regenerated into the water column at a later time. The explicit consideration of sediment release has been included in models by Lorenzen *et al.* (1976) and others. This problem is worth considering in detail here for two primary reasons. First, the coupled lake–sediment system has some dynamic properties that are far different from the lake system alone. Second, the study of the system requires some mathematics that will be convenient to refer to in later sections of the book in discussions of more complex systems. Nonetheless, this section can be skipped by readers without loss of continuity.

Let us refer back to Equation (2.13), which describes solute dynamics in a single compartment, such as a lake, where there is loss to the sediments. Assume now that the sediments are able to release some of the nutrient back into the water. To take this into account, let C_s be the concentration of the nutrient in the sediments $(g\,m^{-3})$, $V_s\,(m^3)$ be the volume of sediments in close-enough contact with the water that the nutrient is exchangeable with the water column, and k_s be the rate constant for release of the bioelement into the water (Figure 2.6). Separate equations are needed to describe C and C_s. As before, determining these equations is simply a matter of applying conservation laws. The equations are, first in words and then in symbols,

(Rate of change of nutrient mass in water) = (input rate due to water inflow) − (loss rate due to water outflow) − (deposition rate to sediments) + (regeneration from sediments)

or

$$V\,dC/dt = I_n - qC - vAC + Ak_sC_s \tag{2.18a}$$

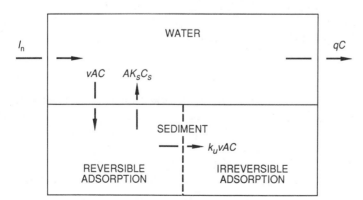

Figure 2.6 Schematic diagram of solute dynamics in a water body that includes exchange of nutrients with sediments.

and

(Rate of change of nutrient in available form in sediment) = (input due to deposition from water column minus rate at which nutrient becomes permanently unavailable) − (losses due to regeneration into water column)

or

$$V_s \, dC_s/dt = vAC(1 - k_u) - Ak_sC_s \qquad (2.18b)$$

where k_u ($0 \leqslant k_u \leqslant 0$) is the fraction of input from the water column to the sediments that becomes permanently inaccessible to any further circulation in the system, due to irreversible adsorption in the sediments or an equivalent process. Note that this loss is assumed to occur at a rate proportional to the concentration of solute, C, in the water.

This pair of coupled equations is linear and thus can be solved analytically. However, the derivation of the solution is complex and is, therefore, omitted. The solution, which, because of its length is shown in full form only in Appendix A (Equations A.1a and A.1b), is identical with that of Lorenzen et al. (1976) with some changes in notation.

The solutions shown in Equations (A.1a and A.1b) for the dynamic behaviour of the system of water column and sediment are too complicated to lead to any quick insights into system dynamics. However, the steady-state solutions are simple enough to comment on directly. Let us first consider these and then discuss some aspects of the transient behaviour.

Steady-state solutions

When the coupled water–sediment system is subjected to the load I_n for a long enough period of time ($t \to \infty$), the transient dynamics decay, leaving only the constant terms in Equations (A.1a and A.1b):

$$C^* = C_1C_5/(C_2C_5 - C_3C_4) = I_n/(q + vk_uA) \qquad (2.19a)$$

$$C_s^* = C_4C_1/(C_2C_5 - C_3C_4) = I_nv(1 - k_u)/[k_s(q + vk_uA)] \qquad (2.19b)$$

Note that C^* in Equation (2.19a) has a form that is very similar to Equation (2.15). The term vk_uA in Equation (2.19a) represents the permanent loss to the sediments, as does vA in Equation (2.15). Reversible exchange with the sediments does not affect C^*. The steady-state solute temporarily stored in sediments, C_s^*, is a multiple of C^* by $v(1 - k_u)/k_s$, the ratio of the fluxes of solute to and from temporary storage in sediments. These equations can be examined in various limits.

Case 1: There is no permanent loss of nutrients from the system (i.e. $k_u = 0$).

Then

$$C^* = I_n/q = C_I$$
$$C_s^* = I_n v/(qk_s) = C_I v/k_s$$

so that the amount of nutrient in the water is the same as if no sediment were present. The effect of increasing k_u from zero is to decrease both C^* and C_s^*.

Case 2: The regeneration rate k_s increases indefinitely.

As $k_s \to \infty$

$$C^* = I_n/(q + vk_u A)$$
$$C_s^* \to 0$$

Although one might intuitively expect an increase in regeneration to cause an increase in C^*, the only effect is a decrease in C_s^*. The reason for this is simply that C^* does not depend on the regeneration rate, only on the loss rate to the sediments.

Case 3: The loss rate to sediment, v, increases indefinitely.

As $v \to \infty$

$$C^* \to 0$$
$$C_s^* \to I_n(1 - k_u)/(k_s k_u A)$$

Hence an unlimited increase in v causes water column solute to go to zero and available sediment concentration to approach a constant. C_s^* does not increase indefinitely because there is a loss to permanent storage in the sediments. However, if $k_u \to 0$, so that the rate of permanent loss increases,

$$C^* = I_n/q$$
$$C_s^* \to \infty$$

that is, the storage of nutrient in the sediment increases indefinitely. The return flux to the water column balances the loss from the water column to the sediment, keeping the concentration in the water column the same as if there were no sediment.

Transient solutions

The behaviour of the transient solutions of Equations (2.18a,b), describing the dynamics of the system from the initial conditions to the steady state, depends on the values of two quantities derived from the equations, the eigenvalues. The single eigenvalue of a one-compartment system was discussed earlier. This two-compartment system has two eigenvalues, λ_1 and

λ_2. Because these are rather complex, they are shown in Appendix A. The fact that there are now two eigenvalues rather than one eigenvalue, as there was in the analyses of the single-compartment models [Equations (2.1) and (2.13)], means that there are two time scales of system behaviour, the inverses of the two eigenvalues. One example will illustrate the importance of these two time scales.

Example: Assume that the rate of bioelement regeneration from sediments, k_s, is small. If k_s/V_s is small enough [compared with $(q + vA)/V$], the two eigenvalues can be shown from Equation (A.2) to be approximately

$$\lambda_1 \cong -(q + vA)/V \tag{2.20a}$$

$$\lambda_2 \cong -(Ak_s/V_s)(q + vk_s A)/(q + vA) \tag{2.20b}$$

where $|\lambda_2| \ll |\lambda_1|$. The effects of these two time scales can be seen by considering the case in which the initial bioelement concentrations (phosphorus in this case) in the water column and sediments, C and C_s, are both at steady-state values; $C^* = 4.03 \times 10^{-6}$ and $C_s^* = 4.8 \times 10^{-2}$ at time $t = 0$, at which point the loading I_n is stopped. Parameter values from Lorenzen *et al.* (1976) are used: $A = 10^8 \text{ m}^2$, $V = 3.8 \times 10^9 \text{ m}^3$, $q = 9 \times 10^8 \text{ m}^3 \text{ year}^{-1}$, $I_n = 45\,000 \text{ kg of P year}^{-1}$, $V_s = 10^7 \text{ m}^3$, $v = 36 \text{ m year}^{-1}$, $k_s = 0.0012 \text{ m}$ year^{-1} and $k_u = 0.6$.

Concentrations C and C_s are plotted on a logarithmic scale in Figure 2.7. Note that there is an initial rapid decrease in C, which largely reflects the rapid washout of phosphorus initially in the water column of the lake. This rate of increase is governed primarily by the eigenvalue with larger absolute value λ_1. This is followed, however, by a much slower decline, as shown by the less steeply sloped second part of the curve in Figure 2.7. The concentration in the lake only slowly reaches its steady-state value, over a longer time scale determined by λ_2. This smaller (or in larger systems the smallest) eigenvalue is often referred to as the critical or dominant eigenvalue, as it governs the long-term dynamics of the system. This is because the phosphorus held in the sediments is slowly released and partially compensates for the loss of external input I_n. It is important to note that the sediments perform a buffering action in the lake (see also the model by Jorgensen *et al.*, 1975). The buffering of the lake is important in two respects. First, when there is a sudden jump in input of solute to the lake, this will be partially compensated for by increased sediment uptake. The second aspect, or reverse side of the coin, however, is that if the solute input is decreased (say, by a reduction in sewage input) this will be offset to some extent by increased net flux from the sediments to the water column for some period of time before the input reduction has an effect.

The models discussed up to now are a vast simplification of actual systems. A real body of water is rarely a single well-mixed compartment, as lakes tend to stratify into layers that increase in density with depth and

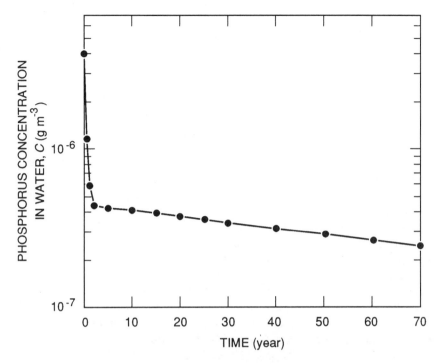

Figure 2.7   The response of the phosphorus concentration in a lake modelled by Equations (2.18a,b) to a cessation of phosphorus input, I_n. There are two distinct parts of the transient behaviour; a sharp drop described by the eigenvalue $\lambda_2 = -3.3$ and a slower decline described by the eigenvalue $\lambda_1 = -0.011$ (see text for discussion).

which do not mix (for very shallow lakes, however, the one-compartment assumption is often a good approximation). It may often be reasonable, however, to partition the lake into just two layers, an epilimnion and a hypolimnion, each of which is relatively well mixed. A number of models of this form exist, including that of Imboden (1974).

I will not consider this complication of the abiotic system. However, it should be remarked that separation of the water column into an epilimnion and hypolimnion (where the epilimnion is the upper layer of warm water in lakes that may form during summer above cooler denser water of the hypolimnion) may be important in models when the hypolimnion is likely to become anaerobic (low in available oxygen). Lakes with anaerobic hypolimnia have smaller values of phosphorus retention than do lakes with aerobic hypolimnia. Phosphorus retention in the hypolimnion often involves complex reaction of the orthophosphate, or dissolved inorganic form of phosphorus, with oxides of iron, manganese and other metals. These form insoluble precipitates under aerobic conditions, but at low levels of dissolved oxygen (or low redox potentials) these precipitates are dissolved (Day *et al.*, 1989). The existence of anaerobic conditions can thus increase the normal release rate of phosphorus (see, for example, Nürnberg, 1984).

2.7 SERIES OF LAKES

Lakes and reservoirs are frequently connected in chains, and an effect on one, whether it is an increase or a decrease in bioelement loading, may propagate to others downstream. Thus, a series of lakes must be considered to be one connected system. It is useful here to consider a model for the propagation of effects in a sequence of lakes, because it is an analogue, in some sense, for the propagation of bioelement effects through a food web (though a very simple one in which there are no feedbacks). A number of models of this type have been constructed. For example, O'Connor and Mueller (1970) modelled chloride concentration in the Great Lakes, represented as a series of boxes connected by flows. Each box receives precipitation and runoff and is subjected to evaporation.

Another system that has been modelled is a system of four shallow eutrophic lakes in Sweden, which Ahlgren (1980) represented schematically (Figure 2.8). A given lake i has a runoff input, $q_i C_{I,i}$, and an input of sewage, J_i (no volume of water is associated with this sewage), a sediment retention factor, R_i, and a concentration, C_i. Therefore, the equations for this system are [generalizing from Equation (2.13) and assuming no evaporative losses, so that water flow accumulates additively],

$$V_1\, dC_1/dt = (q_1 C_{I,1} + J_1)(1 - R_1) - q_1 C_1 \tag{2.21a}$$

$$V_2\, dC_2/dt = (q_1 C_1 + q_2 C_{I,2} + J_2)(1 - R_2) - (q_1 + q_2)C_2 \tag{2.21b}$$

$$V_3\, dC_3/dt = [(q_1 + q_2)C_2 + q_3 C_{I,3} + J_3](1 - R_3)$$
$$- (q_1 + q_2 + q_3)C_3 \tag{2.21c}$$

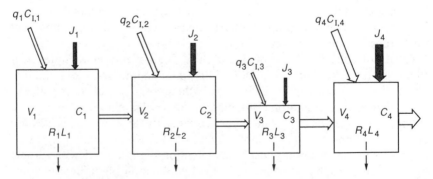

Figure 2.8 Scheme of a series of lakes with solute concentrations C_i, modelled by Ahlgren (1980). The terms $R_i L_i$ represent retention rates in sediments, where the L_i's are inputs to the lakes; e.g. $L_1 = q_1 C_{I,1} + J_1$.

$$V_4 \, dC_4/dt = [(q_1 + q_2 + q_3)C_3 + q_4 C_{1,4} + J_4](1 - R_4)$$
$$- (q_1 + q_2 + q_3 + q_4)C_4 \qquad (2.21d)$$

where the nutrients retained in the sediments are here removed directly from the inflow, and so are not proportional to the concentration in the lake. Thus, the retention factors, R_i, are not expressed by vA_i, as they would otherwise be [in analogy with Equation (2.13)].

A rather remarkable thing about the complex set of equations (2.21) is that they are analytically soluble, and without much difficulty, since they form a 'cascade' sequence with no feedbacks. One can solve for the behaviour of the concentration C_1 in the highest lake in the drainage first, and use the result as part of the input to the next lake below it in the sequence, and so forth. To illustrate the method, while avoiding complexity, we compute only the first two lake concentrations, C_1 and C_2. The solutions are shown in Appendix B for a special case where the initial concentrations of all the lakes are equal to zero. It is clear that the solutions of successive lake concentrations become more and more complicated. However, formulae for variables at any stage in the sequence of this type of cascade are available (e.g. Bailey, 1964).

Ahlgren (1980) used the data in Table 2.2 to specify the Equations (2.21a,b,c,d) for phosphorus. The different values of $C_{1,i}$ reflect different land usages around the lakes, with Lake no. 4 surrounded by a relatively greater amount of farmland. The water discharge values, q_i, were found by multiplying specific runoff values per unit area of land by the sizes of the drainage areas, A_i.

Until 1970, all four lakes received domestic and industrial sewage. In 1970 the complete diversion of all sewage was begun. Measurements of phosphorus and nitrogen in the lakes were begun at that point. To compare the data with theory, computer simulations were performed by Ahlgren (1980) on Equations (2.21). In Figure 2.9 comparisons are made between simulation results and data for phosphorus concentrations after 1970, both including and ignoring retention. Although a retention of $R = 0.44$ was expected, based on relevant studies of similar lakes, the assumption of zero

Table 2.2 Data used for the parameter values of Equations (2.21) (Based on data in Ahlgren, 1980)

Lake no....	1	2	3	4
Drainage area, A (km^2)	50.7	94.0	128	264
Lake area, A_1 (km^2)	6.1	2.67	1.0	1.6
Lake volume, V (10^6 m^3)	15.4	14.3	3.0	5.3
Runoff, q (l s^{-1} km^{-2})*	5.2	5.2	5.2	5.2
Runoff P concentration, D (g m^{-3})	0.1	0.05	0.05	0.05
Sewage P loading (g m^{-2} year^{-1})	0.16	0.5	2.2	2.1

*1978 value.

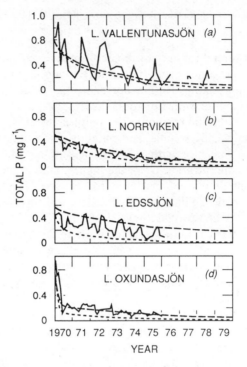

Figure 2.9 Comparisons of data on total phosphorus concentration in four Swedish lakes (Figure 2.8) after diversion of sewage effluents (solid lines) with model simulations (dotted and dashed lines). In the dashed model curves, the retention $R_p = 0$, while in the dotted curves $R_p = 0.44$. (From Ahlgren, 1980.)

retention worked better in the model. Ahlgren attributed this to excessive phosphorus release from the sediments. Ahlgren also applied the model to nitrogen, but the results were less satisfactory, probably due to denitrification losses or less accurate input data.

The derivation of this cascade solution for solute in a series of lakes is an appropriate place to introduce an additional stability concept, that of resistance.

Resistance

Resistance can be defined in a manner similar to Harrison and Fekete (1980) as the relative magnitude of the change in a specified component of system following a unit change in one of the fluxes, either an input, an output, or a flow between compartments in the system.

The concept of resistance is related to the sensitivity analysis that modellers frequently undertake to see how variables in their model respond to changes in inputs, fluxes, rate coefficients or other quantities in the model. There is no one overall resistance of a system, but a variety of different resistances associated with changes in various fluxes and the resultant deviations in compartments of interest. Taking Ahlgren's lakes as an example, suppose there is a sudden increase in phosphorus input in sewage, J_1, so that the total input of sewage to the first lake, J_1, increases by a certain amount. What is the effect on the concentrations, C_1, C_2, C_3 and C_4 of the four lakes? From Ahlgren's parameter values, these values were computed through time following an increase in J_1 (Figure 2.10). One can compute the percentages of the final changes of C_1, C_2, C_3 and C_4 relative to the fractional change of J_1 (which is 0.1), or

$$\text{Fractional change in } C_i = [(C_i^{**} - C_i^*)/C_i^*] \qquad (2.22)$$

where C_i^* is the steady-state value of C_i before the perturbation and C_i^{**} is the steady-state value after the perturbation. The resistance will be defined here as the inverse of this fractional change in C_i, so that resistance increases as this relative fractional change decreases. The computed resistances for the four lakes, if expressed as the inverses of the fractional changes in C_i for the unit change in J_1, were approximately Resistance$_1$ = 1.0834, Resistance$_2$ = 2.522, Resistance$_3$ = 5.866 and Resistance$_4$ = 8.510, where greater values denote greater resistance to change. It is clear that in this case the

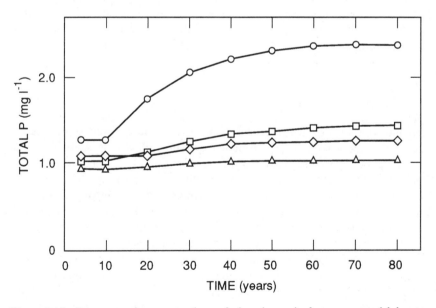

Figure 2.10 Response of concentrations of phosphorus in four connected lakes to a 100% increase in sewage input to Lake 1 in year 10: Lake 1 (circles), Lake 2 (squares), Lake 3 (diamonds) and Lake 4 (triangles). [Based on Ahlgren's (1980) model, Equation (2.22).]

compartments further removed from the perturbation have the greatest resistance.

Resistance can also be thought of as having a temporal component; that is, not only is the magnitude of change important, but also how long it takes that change to occur. In the present case. Resistance$_2$ = 74.56 at 3 years and 57.27 after 10 years. Clearly, on a short time scale, the resistance to change is high.

The cascading compartment model is a special case of a more general model in which compartments are linked sequentially, but where flows can occur in either direction. Solution of the equations of such a model is much more difficult than solution of the cascade model equations, but the more general model has wider applicability. It has been used for such purposes as the modelling of the nutrient dynamics of connected ocean basins. Wulff (1989) for example, computed the temporal nutrient budgets in the Baltic Sea, and Sarmiento *et al.* (1988) examined the causes of anoxia (severe reduction in oxygen) in the Mediterranean Sea with such models.

The models discussed above are idealizations of real systems. Input flows of water and concentrations of bioelements usually undergo significant temporal fluctuations, as do outflows and the amount of water in the lake. Evaporation and direct precipitation also may have significant effects. While the simple models analysed above fail to take into account these complicating factors, it is possible to derive more general models, based on conservation laws, to represent systems of almost any degree of complexity. Computations of nutrient loading, for example, have been carried out for highly complex spatially heterogeneous and temporally varying cases (e.g. Huff *et al.*, 1973; Bolla and Kutas, 1984).

2.8 SUMMARY AND CONCLUSIONS

The purpose of this chapter has been to introduce some of the basic concepts of flow of solutes through a system and, at the same time, to introduce the types of stability that are most frequently applied in studies of food webs. The systems analysed here are all simple linear systems, and also 'passive' ones in the sense that the compartments do not actively affect their inflows or outflows, Such systems are appropriate starting places for development of concepts and models, as they do not exhibit some of the really surprising features of systems with biotic elements.

The single-compartment model of a nutrient in water solution subjected to an external loading and losses that are proportional to concentration illustrated the principles of steady state and transient components of the solution. The linear system analysed was both locally and globally stable; that is, it recovered to steady state from all plausible perturbations. The rate of recovery is termed resilience and identified with the eigenvalue of the system or the inverse of the return time to equilibrium. In linear systems this return time is equal to the turnover time or residence time of nutrient in the compartment.

Nutrient dynamics in real systems, even in the absence of biota, are much more complex than can be represented by a single-compartment model. There are interactions of the nutrient solute with sediments or solid components of clay and soil. Coupled equations are required to describe the nutrient dynamics between the aqueous and solid phases. Losses to sediments, both temporary and permanent, can affect the dynamics of the nutrient in the water body. A series of connected compartments, allowing either one-way or two-way flow, may be necessary to describe the nutrient dynamics of many systems that are not internally homogeneous. Another aspect of stability, resistance, is used to measure the inverse of the amount of change of a given compartment to changes in one or more of the flux rates in the system. The speed of change is also important. A system that appears to be highly resistant over a short time period may undergo great changes in response to flux rate changes over a long time period.

One of the useful results shown so far is the equality of mean residence time, T_{res}, a key concept of nutrient flow, and return time of a system to steady state, T_R, an important stability concept. This equality will be shown to be at least a good approximation in the more complex systems studied in later chapters.

3   Nutrients and autotrophs

3.1   INTRODUCTION

The inclusion of living components in models of nutrient kinetics changes the nature of these models from the simple systems reviewed in Chapter 2, in which compartments were passive recipients of nutrient molecules and had no control over their rate of input. Living organisms grow and actively incorporate energy and nutrients into their own biomass and biomass of their progeny. They acquire these nutrients by either uptake from the abiotic environment or consumption from the biomass of other organisms in the food web. The movement of nutrients in food webs thus becomes largely determined by biological and ecological processes, which can be far more complex than the geochemical processes that otherwise control the movements of material elements.

Autotrophs, or primary producers, are at the base of the food web and acquire most of their energy and nutrients from abiotic sources (though there exist some autotrophs, such as carnivorous plants, that supplement their abiotic nutrient supply by capturing and consuming small animals). However, the methods of nutrient acquisition differ somewhat across autotroph types. Phytoplankton cells, for example, are immersed directly in the water medium that supplies their nutrient needs. Nutrient ions, such as nitrate (NO_3^-), ammonium (NH_4^+), phosphate (PO_4^{4-}, HPO_4^{3-} or $H_2PO_4^{2-}$, depending on pH), and sulphate (SO_4^{4-}), are transported from the external medium to the interior of the cell by specialized enzymes called permeases. In the higher terrestrial plants the root systems, which are in contact with soil water, are the primary means of acquiring needed mineral nutrients, while carbon is largely taken up through stomata on the leaves. Then the nutrients, through diffusion or, in higher plants, translocation, move to sites (usually in the roots and leaves of higher plants) where they are assimilated into organic compounds needed for growth and regulatory functions. Plants store the nutrients for various periods of time in their biomass, eventually releasing these nutrients through processes of excretion and organism death and decomposition.

The biological processes of growth and regulation are dependent on a sufficient supply of nutrients. The primary atomic components of biomass are carbon (C), hydrogen (H), oxygen (O), nitrogen (N), phosphorus (P), sulphur (S), calcium (Ca), potassium (K) and magnesium (Mg). The first three of these elements make up carbohydrates, which constitute most of the dry weight of plant tissues. The latter six are needed in smaller, but still relatively large concentrations, and so are referred to as macronutrients. Among other things, N is a major component of proteins, nucleic acids and chlorophyll, P is found in adenosine triphosphate (ATP), nucleic acids and cell membranes, S occurs in proteins, K actuates some enzymes and plays a role in plant water balance, and Mg is a component of chlorophyll. Also essential to living tissues are sodium (Na), chlorine (Cl), and a number of micronutrients in percentages less than 0.05%: iron (Fe), manganese (Mn), zinc (Zn), iodine (I), boron (B), silicon (Si), copper (Cu), molybdenum (Mo) and cobalt (Co).

As introduced in Chapter 1, stoichiometry refers to the ratios of chemical constituents of matter. In living biomass some generalizations concerning stoichiometry are possible. A frequently observed set of uptake ratios for carbon, nitrogen and phosphorus by ocean plankton is (by atom) 106 C:16 N:1 P or (by weight) 42 C:7 N:1P, which is known as the Redfield ratio. This figure is certainly not fixed, however, and Boyd and Lawrence (1966) found freshwater phytoplankton with ratios (by weight) of 116 C:9.9 N:1P. The ratios of the six major elements (by atom) in plankton have been estimated by Stumm and Morgan (1981) and Trudinger et al. (1979) to be 106 C:263 H:110 O:16 N:1 P:0.7 S.

An equivalent of the Redfield ratio is seldom given for terrestrial plants, as the ratios tend to vary considerably among species due to differences in the support structures. However, Attiwill (1980) calculated concentrations of several macronutrients in green foliage and litter in various forest types. For example, the concentrations of these macronutrients (in g kg^{-1} dry weight) in green foliage are shown in Table 3.1. Such concentrations are highly variable and change continually (e.g. Fife and Nambiar, 1982).

The existence of rather well-defined, if often variable, ratios of bioelements in green plant (referred to equivalently as primary producer or autotroph) biomass means that the nutrient needs of autotrophs will best be satisfied when the amounts of nutrients available in the surrounding medium occur in approximately the same proportions as the proportions of these elements in the biomass. When the proportion of one particular nutrient to the other nutrients available in the environment is below that needed for plants or

Table 3.1 Concentrations (g kg^{-1} dry weight) of five macronutrients in three forest types (from Attiwill, 1980)

Forest type	N	P	K	Ca	Mg
1. Australian eucalypts (wet sclerophyll)	14.73	1.05	4.60	6.77	3.87
2. Mixed Quercus, Betula and Fraxinus	24.41	1.32	11.08	11.94	2.71
3. Equatorial forests	25.92	1.62	12.72	18.99	1.73

animals to grow and perform biological functions, it may act as a limiting factor in biomass growth, as discussed in the next section.

3.2 BIOMASS GROWTH AND LIMITING NUTRIENTS

To grow successfully, green plants need physical space, solar radiation, water, carbon dioxide and sufficient amounts of all of the macronutrients and micronutrients. Any of these factors can become limiting to plants, but of most interest here are situations where nutrients are factors that are or can be limiting.

In general, plants are adapted to their environments through physiological mechanisms that balance their growth demand for various nutrients to the supply from the surroundings. Since environmental conditions are highly variable, however, it is unlikely this balancing will be perfect.

Justus Liebig, a pioneer in the study of various factors affecting the growth of plants, had the insight that the growth rate would be regulated by the 'foodstuff' obtainable in smallest amount relative to its needs. This idea is now called Liebig's law of the minimum (e.g. Odum, 1971, p. 106). One qualification on this principle is that it is strictly applicable only under steady-state conditions. Even under those conditions, it is unusual for the limitation on growth to be governed by only one limiting nutrient. It is more likely that when two or more nutrients are in short supply, each contributes to some degree in limiting growth, not only the nutrient that is 'most limiting'. This has been observed, for example, in studies in arid grasslands, where increases in either water or nitrogen improve growth rates of some species (Lauenroth, 1979). Pot and chemostat experiments have documented numerous cases where at least two nutrients were simultaneously limiting (e.g. Droop, 1974; Gates and Wilson, 1974; Shaver and Melillo, 1984).

The implications of multiple limiting nutrients will be discussed in Chapter 9. Throughout most of this book, however, the analysis of the effects of limitation on food web dynamics will be restricted to cases of only one potentially limiting nutrient. If this nutrient is abundantly available, then the growth rate of the autotroph will be assumed to approach a maximum value that is, for present purposes, a constant depending on a combination of the physiological characteristics of the organism that limits its processing rates and factors such as physical space, water and solar radiation.

In principle, every one of the essential elements can be limiting and, no doubt, examples of limitation by a given nutrient can be found somewhere on earth, especially in terrestrial systems, which are more heterogeneous by nature than are aquatic and marine systems. However, in a very large number of cases either N or P appear to be limiting. Table 3.2 lists a variety of studies of different ecosystem types and the nutrient(s) determined to be limiting in each study. The analyses of this book are, for the most part,

Table 3.2 Some studies in which nutrient limitations in natural systems were discussed. The * means that no particular nutrient, but nutrient limitation in general is discussed and ** means that nutrient limitation in a wide variety of ecosystems is discussed

Paper	Nutrient(s)	Type of ecosystem
Beadle (1954)	P	Dry sclerophyllous forests of southeastern Australia
Penning de Vries et al. (1974)	N	Loblolly pines in southeastern US
Weinstein et al. (1982)	N,Ca	Temperate forests
Jordan and Herrera (1981)		Tropical forests
Kliejunas and Ko (1974)	N,P	'Ohi' a forest in Hawaii
Tanner (1985)	N,K,P,Ca	Tropical mor ridge forest
Golkin and Ewel (1984)	P	Forests and lakes in Florida
Reuss and Innis (1977)	N	Temperate grassland
Black and Wight (1979)	N,P	Northern great plains
Shaver and Chapin (1980	N,P	Cotton grass-tussock tundra
Kachi and Hirose (1983)	N	Coastal sand dune soils
Hopkinson and Shubauer (1984)	N	Salt Marsh
Howard-Williams and Allanson (1981)	P	Littoral zone of South African lake
Boynton et al. (1982)	N,P	Estuarine phytoplankton
Kalff (1983)	P	Algal biomass in tropical lakes
Schindler et al. (1971), Schindler (1977)	P	Canadian shield lakes
Robarts and Southall (1977)	N	Algal growth in reservoirs
Morris and Lewis (1988)	N,P	Phytoplankton in mountain lakes
Caraco et al. (1987)	N,P	Phytoplankton in coastal brackish ponds
Peterson et al. (1985)	P	Algae in a tundra river
Smith (1979)	P	Phytoplankton in lakes
Sakshaug and Olsen (1986)	N,P	Phytoplankton in coastal waters (fresh to marine)
Hecky and Kilham (1988)	N,P	Phytoplankton in fresh to marine water
Smith (1984)	N,P	Phytoplankton in marine ecosystems
Howarth (1988)	N,P	Marine ecosystems
Steele (1962)	*	Marine ecosystems
Harvey (1963)	*	Marine ecosystems
Riley (1963)	*	Marine ecosystems
Dugdale (1967)	*	Marine ecosystems
Gutschick (1981)	N	**
Lovstad (1984)	P,N	**
Agren (1985)	N	**

general in the sense that no particular nutrient is assigned the role of limiting nutrient. The results should apply equally to a variety of potentially limiting nutrients. Whenever the term 'nutrient' is used, it will generally refer to a nutrient that is potentially limiting.

The concept of a limiting nutrient is relatively simple in the case where the nutrient availability is completely controlled externally by geochemical processes. In actuality, however, the autotrophs and other organisms in an ecosystem also exert their own degree of control over nutrient availability by forming internal nutrient cycles in which nutrients are continually released from biomass back into the available pool to be used again. Available nutrient is thus composed of both 'new' and 'regenerated' or 'recycled' nutrients, the latter being determined by the ecological system as a whole.

Switzer and Nelson (1972) identified three levels of the circulation of nutrients in forest ecosystem: geochemical, biogeochemical and biochemical (Figure 3.1). Geochemical cycles encompass inputs from and losses to the larger abiotic environment. Biogeochemical cycles encompass the recycling of nutrients between the soil and plants and heterotrophs. Biochemical cycles encompass internal transfers or translocations of nutrients within plants. There are major differences in the degree to which these cycles govern the movement of different nutrients. The need for phosphorus in tree metabolism, for example, is met largely through internal translocation, with only about 5% coming as new input from the geochemical cycle (the left-hand side of Figure 3.1). Interesting differences in the timing of translocation of nutrients between different tree types have been noted. For example, the seasonal pattern of translocation of phosphorus and nitrogen between leaves, stems and roots differs between tundra deciduous and

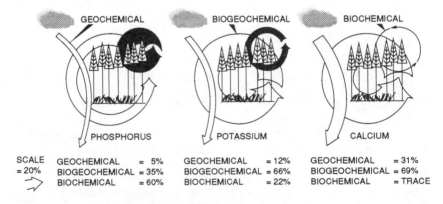

SCALE	GEOCHEMICAL	= 5%	GEOCHEMICAL	= 12%	GEOCHEMICAL	= 31%
= 20%	BIOGEOCHEMICAL	= 35%	BIOGEOCHEMICAL	= 66%	BIOGEOCHEMICAL	= 69%
	BIOCHEMICAL	= 60%	BIOCHEMICAL	= 22%	BIOCHEMICAL	= TRACE

Figure 3.1 A comparison of the contributions of geochemical, biochemical and biogeochemical cycles to the phosphorus, potassium and calcium requirements of a 20-year-old loblolly pine plantation in southeastern US. (From Switzer and Nelson, 1972.)

evergreen shrubs (Chapin *et al.*, 1980). In deciduous shrubs, up to 14% of the total plant N and P was translocated from stems and large roots to leaves during the three weeks following snow-melt. In contrast, evergreen shrubs translocate N and P to leaves gradually through the growing season. Stems and large roots do not appear to act as large nutrient stores.

The plant's K needs are met predominantly through uptake from the soil, where it is the product of decomposition of plant detritus. The proportion of Ca used by the plant that is new input from the geochemical cycle is much greater than that of either P or K. This results from the great accumulation of Ca in the permanent tissue of plants, whereas K and P are primarily used in the more transient biomass and regulation processes of foliage where turnover rates are much higher.

Biogeochemical recycling may also be important in aquatic and marine ecosystems, especially oligotrophic systems. Dugdale and Goering (1967), Dugdale (1967) and others have distinguished between new nutrients supplied externally to the photic zone of marine ecosystems from outside through eddy diffusion, and recirculated nutrients within the photic zone. The latter may account for up to 94% of production in oligotrophic systems. As a final example, in dense aquatic macrophyte beds (*Potamogeton pectinatus L.*) the movement of phosphorus has been shown to be almost a closed cycle, in which the phosphorus released from decaying macrophytes was rapidly taken up again by epiphytic algae (Howard-Williams and Allanson, 1981).

It is not surprising that plants have special adaptations to help recover and thus conserve nutrients that are limited in their rate of supply by the geochemical cycle. Particularly important for terrestrial plants on nutrient-poor soils are extensive root systems concentrated near the soil surface (Jordan and Herrera, 1981). This reduces the likelihood of loss of nutrients through leaching and helps the plants compete with decomposers for nutrients. In tropical forests, aerial roots are another adaptation. These roots can penetrate and obtain nutrients from mats of dead epiphytes, such as bromeliads.

Headley *et al.* (1985) discussed a number of strategies that tundra evergreen plants, such as the moss *Lycopodium annotinum*, use to recycle nutrients internally and to improve the efficiency of nutrient uptake from the soil, and thus to conserve nutrients. These include increased root:shoot ratios and surface areas of roots, internal recycling from old to new tissues, efficient use of nutrient at low internal concentrations and a temporal extension of the period of photosynthetic activity through the evergreen habit.

A highly effective adaptation of virtually all forests is the symbiosis between trees and mycorrhizal fungi. The hyphae of these fungi are efficient at taking up nutrient from decomposing litter (they may actually participate in decomposition; Janos, 1983). They pass much of these absorbed nutrients on to the plant roots with which they are associated, receiving some carbohydrates from the plant in return. This symbiosis enhances nutrient uptake by the plants.

Using information from a variety of forest ecosystems organized into the form of linear flow models, Finn (1976, 1978, 1980) calculated the cycling indices for various nutrients. The cycling index represents the recycled atoms as a proportion of the total nutrient throughput. This ratio for forest ecosystems ranged between 0.6 and 0.8. The cycling index will be discussed in more detail later in this chapter.

Kelly and Levin (1986) reviewed data from more than 200 natural areas on the relative importance of internal cycling and external sources. They noted patterns of decreasing nutrient use efficiency as a function of the amount of external loading. Whether or not there is nutrient limitation under these circumstances is hard to determine. The significance of the nutrient limitation can only be understood in the context of the complete system. The next few sections study the effects of nutrient limitation through formulation and analysis of simple mathematical models for systems in which the autotroph is the highest trophic level.

3.3   FORMULATING A MODEL OF NUTRIENT CYCLING

As Dugdale (1967) pointed out, the importance of nutrient limitation lies in its effects on the dynamic behaviour of systems. It can equally be said that the nutrient limitation is to some extent a function of the dynamics of the system. Both aspects will emerge from a model analysis of a system with nutrient cycling.

In this and the following chapter the only living organisms considered are primary producers. To simplify matters for the present, all autotrophic organisms are lumped into one compartment, so that autotophic biomass can be represented by a single variable. This is a reasonable representation when all the autotrophs in the system are members of the same species. When a mixture of different species is considered, the validity of aggregation is questionable, but the advantage of simplification to model analysis is so great that the approach seems justified, at least as a first step. In many cases a simple model will capture the essence of a phenomenon. Disaggregation into competing autotrophic species will be considered in Chapter 9.

There are two general ways of modelling the dynamics of nutrient-limited autotrophs, depending on whether nutrient uptake by the autotroph is assumed to be controlled solely by the external levels of available nutrient, or whether the intracellular nutrient level of the autotroph also exerts some control. Modellers assuming the former type of control usually assume that growth is described by a Monod-type function (e.g. Riley et al., 1949; DiToro et al., 1971, 1977; Thomann et al., 1974). The second model type is often referred to as the cell quota model (e.g. Droop, 1968; Caperon, 1968). By including a mechanism for the effect of internal nutrient levels, such models may be more appropriate for looking at details that require separation of the nutrient uptake and primary production processes. However, there is evidence that in many situations the simpler assumption of only external nutrient control of autotroph growth rate is an adequate approximation of growth dependence on nutrients (e.g. Caperon, 1965; Eppley and Thomas, 1969; Golterman et al., 1969; O'Brien, 1974; Button, 1978; DiToro, 1980; Auer et al., 1986). Because of the analytical simplification afforded by this assumption, it will be made throughout this chapter. The implications of the cell quota type of model will be explored in Chapter 4.

Both of the model types discussed above were derived for phytoplankton or bacteria and so neither can be expected to be appropriate for higher plants, where tissue differentiation and complex internal storage and translocation processes can occur. However, because models that lack internal differentiation may still adequately describe plant growth over long time scales, these models are frequently used to describe growth in higher plants, and will be used here also. The use of models with greater internal differentiation for higher plants will be discussed in Chapter 4.

In the present chapter the nutrient uptake per unit biomass of autotroph and the growth rate of the autotroph are assumed to be controlled solely by the external nutrient level or 'pool'. In addition, the amount of the nutrient in a given quantity of autotroph biomass, X, is assumed to be a fixed mean fraction γ of the biomass, so that the amount in the biomass is γX. Thus, an increase in autotroph biomass X must be accompanied by an uptake of an amount of nutrient γX from the nutrient pool.

Biomass is a convenient variable to use for modelling the flows of energy and matter in an ecosystem. Let $X(t)$ represent the instantaneous autotroph either as a density such as biomass (in $g\,m^{-2}$ or $g\,m^{-3}$, or an amount or standing stock such as g or kg, in a given system) in the primary producer trophic level. The most general possible model for the changes in $X(t)$ through time is

$$dX/dt = (\text{Production of } X) - (\text{Loss of } X) \tag{3.1}$$

where X is used as shorthand for $X(t)$. Here 'Production of X' includes both the growth in biomass of individual autotrophic organisms and the reproduction of new individuals, and 'Loss of X' encompasses both the decreases in biomass of individuals through processes such as respiration, litterfall and biomass sloughing as well as losses of whole individuals through mortality and emigration (by drift of phytoplankton in a lake or washout of periphyton and macrophytes in a stream, for example). For the purposes of this chapter, a special form of Equation (3.1) is used; that is,

$$dX/dt = r(N)X - (d_1 + e_1)X \tag{3.2}$$

where

$r(N)$ = the intrinsic rate of growth of X, which depends on the amount of available nutrient, N.

$d_1 =$ the rate coefficient for the loss of autotroph biomass, where the biomass goes to a detrital compartment but stays in the system (i.e. the nutrient molecules can be recirculated back to living biomass);

$e_1 =$ the rate coefficient for the loss of autotroph biomass, where the biomass is completely removed from the system, by harvesting or washout, for example, so that the nutrient molecules are lost forever from the system.

Input of biomass through autotroph immigration (transport of phytoplankton in water flow into the system, for example) has also been omitted for simplicity, though the term e_1 can be thought of as including, among other types of flux, the net difference between emigration and immigration.

The biomass of the detritus component will be represented by $D(t)$. It will be assumed that the same mean ratio of nutrient weight to biomass weight, γ, exists in detritus as in autotroph biomass. The consequences of this assumption will be examined in Chapter 7. The dynamics of the detrital biomass can then be described by

$$dD/dt = d_1 X - (d_D + e_D)D \tag{3.3}$$

where

$d_D =$ rate coefficient for decomposition of detritus, releasing nutrients into the available pool;

$e_D =$ loss of detritus from the system, due to burial or transport from the system.

The possible complications that may arise from explicit inclusion of a community of decomposers (e.g. bacteria and fungi), such as immobilization of available nutrients, are put off until Chapter 7. In Chapter 2, the equations of solute dynamics [e.g. Equation (2.2)] described the solute in terms of concentration C (say, g solute l^{-1} water). This is an inconvenient variable for present purposes, as we would like to have a more flexible measure of nutrients that is more compatible with biomass X, which may have a variety of units (either a density such as $g\,m^{-2}$, $g\,m^{-3}$, $kg\,ha^{-1}$, or amount or standing stock such as g or kg, in a given system depending on the situation). Therefore, the variable $N(t)$, or N for short, will be used from now on and will have units compatible with whatever units X has. The variable N is related to C through

$$C = L_C N \tag{3.4}$$

where L_C is a factor that can convert N from whatever its units are to g solute l^{-1} water. N represents the 'pool' of limiting nutrient in whatever unit of volume or area is chosen for the model system. If we assume homogeneity within these effective volumes, N should be able to play the same role as concentration, aside from a constant factor.

A conservation equation for N can easily be derived. In the absence of biotic uptake and release, N is described by

$$dN/dt = I_n - r_n N \qquad (3.5)$$

where

I_n = rate of input of nutrient; for example, in $g\,s^{-1}$.
r_n = rate coefficient of output of nutrient in s^{-1}.

Note that I_n refers to input via the geochemical cycle; that is the input of new nutrient. Equation (3.5) is precisely analogous to Equation (2.2), except that in Equation 2.2, C describes grams of the nutrient per unit volume of water, while in the present case N means grams in the system of interest.

When biotic uptake by the autotroph and release by the detritus are included, the equation for N takes the form,

$$dN/dt = I_n - r_n N - \gamma r(N)X + \gamma d_D D \qquad (3.6)$$

Figure 3.2 illustrates the arrangement of compartments. Recall that γ is the ratio of nutrient to biomass (e.g. $g\,g^{-1}$).

The remaining question in specifying a complete model is how to describe the effect of available nutrient on the growth rate, $r(N)$. One possibility is simply to let $r(N) = r_1 N$, where r_1 is a constant. This assumes there is no limit on the rate of biomass production, as long as nutrient is available in sufficient supply. This is an unrealistic assumption since the genetic and environmental limits on growth will ultimately prevent further increases in $r(N)$. A description frequently used is an analogue of the Monod function, which was used initially to describe bacterial growth, and which is itself based on the Michaelis–Menten function for enzyme kinetics. In the present

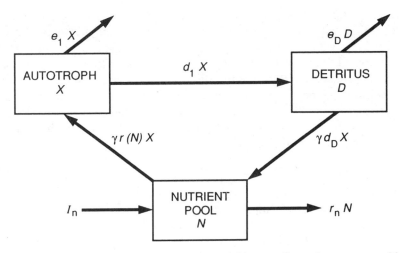

Figure 3.2 Schematic diagram of nutrient and biomass fluxes in a system with a nutrient pool, autotroph and detritus, as described by Equations (3.8a,b,c).

application, this function can be written as

$$r(N) = r_1 N/(k_1 + N) \tag{3.7}$$

The use of this function to describe the growth of phytoplankton was pioneered in the 1960s by, among others, Dugdale (1967), Caperon (1965), Eppley and Thomas (1969) and Golterman *et al.* (1969). This function should not be thought of as a 'law' in any sense, but as an empirical relation that often provides a good fit to data. The parameters can be described as follows:

r_1 = maximum rate of biomass growth (units of $g\,s^{-1}$)
k_1 = half-saturation concentration; that is, the concentration of nutrient at which the rate of biomass production is one-half the maximum rate of production (units of g).

The complete set of equations describing the system is thus

$$dN/dt = I_n - r_n N - [\gamma r_1 NX/(k_1 + N)] + \gamma d_D \tag{3.8a}$$

$$dX/dt = r_1 NX/(k_1 + N) - (d_1 + e_1)X \tag{3.8b}$$

$$dD/dt = d_1 X - (d_D + e_D)D \tag{3.8c}$$

These equations will form the basis for a study of the behaviour of the steady-state values, the local stability, the resilience, the ratio of new production to production based on regenerated nutrients and other properties of the system.

3.4   STEADY-STATE CONDITIONS

The steady-state solutions of Equations (3.8a,b,c) are of particular interest, because they will allow us to predict values of autotroph biomass from nutrient input values, a prediction of importance in the management of real systems, whether one wants to estimate the beneficial effects of fertilization on forest production, the harmful effects of nutrient loading in causing eutrophication of bodies of water or any number of related problems. The determination of the steady-state values is found easily by setting $dN/dt = dX/dt = dD/dt = 0$ and solving the right-hand sides for the steady-state values N^*, X^* and D^*. The results are

$$N^* = k_1(e_1 + d_1)/(r_1 - d_1 - e_1) \tag{3.9a}$$

$$X^* = (I_n - r_n N^*)(d_D + e_D)/[\gamma(d_1 e_D + d_D e_1 + e_1 e_D)]$$
$$= (I_n - r_n N^*)/(\gamma e_1) \quad \text{(if } e_D = 0) \tag{3.9b}$$

$$D^* = d_1 X^*/(d_D + e_D) \tag{3.9c}$$

Two important features of the steady-state values are that N^* is a constant with respect to changes in I_n, whereas X^* increases linearly with increasing I_n. This is exhibited in Figure 3.3 for values shown in Table 3.3. A better understanding of the steady state can be achieved by the detailed study of an abbreviated set of equations, leaving out the detritus (which is merely a passive, linear component in this model, anyway) and shunting

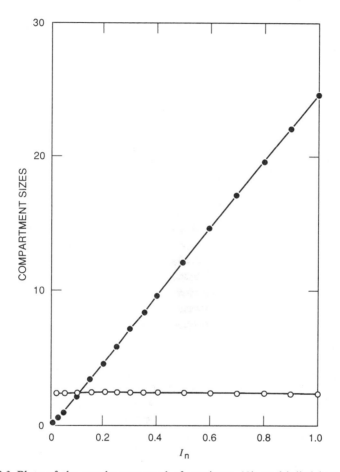

Figure 3.3 Plots of the steady-state pool of nutrients, N^*, multiplied here by 10^3 (open circles) and autotroph biomass (closed circles) as functions of nutrient input, I_n (multiplied here by 10^3). Based on parameter values in Table 3.3.

Table 3.3 Parameter values used in Equation 3.9 for producing the simulation results shown in Figure 3.3. These are hypothetical values and do not relate to any particular system

$I_n = 0.00005$ to 0.001	$d_1 = 0.1$	$e_1 = 0.001$
$r_1 = 0.3$	$d_D = 0.1$	$e_D = 0.001$
$r_n = 0.005$	$\gamma = 0.02$	$k_1 = 0.005$

autotroph losses directly to nutrient regeneration. This is completely equivalent to assuming that the detritus decomposes instantaneously. The condensed model is

$$dN/dt = I_n - r_n N - r_1 NX/(k_1 + N) + d_1 X \qquad (3.10a)$$

$$dX/dt = r_1 NX/(k_1 + N) - (d_1 + e_1)X \qquad (3.10b)$$

The characteristics of this system can be explored by plotting the zero isoclines, obtained by setting $dN/dt = 0$ and $dX/dt = 0$. These are, respectively

$$X = (I_n - r_n N)(k_1 + N)/[(r_1 N - d_1 N) - k_1 d_1] \qquad (3.11a)$$

$$N = k_1(e_1 + d_1)/(r_1 - e_1 - d_1) \qquad (3.11b)$$

These isoclines are plotted in Figure 3.4.

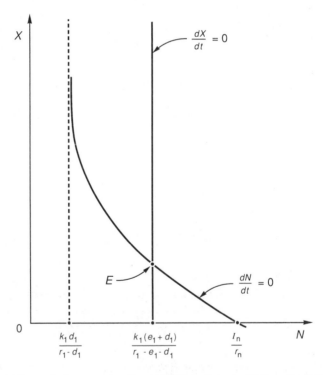

Figure 3.4 Plots of the zero isoclines, $dN/dt = 0$ and $dX/dt = 0$ of the pair of Equations (3.10a,b). The point of intersection of the two isoclines, E, is the equilibrium point of this system (N^*, X^*). The point $N = I_n/r_n$ is the nutrient level in the system described by Equations (3.8a–c) when autotrophs are absent and the point $N = k_1 d_1/(r_1 - d_1)$ is the nutrient level when autotrophs are present and the limit of complete nutrient recycling is approached.

Two properties of the set of equations (3.10a,b) are made clear by Figure 3.4:

1. The possible range of steady-state values of available nutrient, N^*, assuming that there is a non-zero autotroph biomass in the system, lies between $k_1 d_1/(r_1 - d_1)$ and I_n/r_n, as indicated along the N axis in Figure 3.4. When there is no autotroph present ($X^* = 0$), then $N^* = I_n/r_n$. When the autotroph is present, then $N^* = k_1(e_1 + d_1)/(r_1 - d_1 - e_1)$ and the smaller the loss rate e_1, the smaller N^* is.
2. The autotroph will go to extinction if either

$$r_1 < e_1 + d_1 \tag{3.12a}$$

or

$$k_1(e_1 + d_1)/(r_1 - e_1 - d_1) > I_n/r_n \tag{3.12b}$$

An explanation at an intuitive level is possible for both of these properties. The expression (3.12a) implies that no matter how large the available nutrient pool N is, the loss rate of the autotroph biomass may exceed its production, leading to extinction. This can be seen from the right-hand side of Equation (3.10b). Inequality (3.12b) is a more stringent condition on survival. This is equivalent to the fact that if production, $r_1 N/(k_1 + N)$, is less than losses, $(d_1 + e_1)$, even when N takes on its largest achievable steady-state value in the present system, which is I_n/r_n, the autotroph must necessarily go to extinction; that is, autotrophic production must exceed losses when the available nutrient is at its highest reachable steady-state value.

Property 1 can also be interpreted. First, it is impossible for N^* to exceed I_n/r_n because, from Equation (3.10b), $d_1 X^*$ must be less than or equal to $r_1 N^* X^*/(k_1 + N^*)$, so that, from Equation (3.10a), N^* must always be less than or equal to I_n/r_n. The behaviour near the lower limit of possible values of N^*, $k_1 d_1/(r_1 - d_1)$, is more curious. At first it seems strange that decreases in the loss rate of autotroph biomass from the system, e_1, should cause decreases in N^* and, simultaneously, cause X^* to increase towards infinity. To understand this, we should first recall that N^* is entirely controlled by the coefficients of the autotroph: k_1, e_1, d_1 and r_1. The input I_n has no effect on N^*. However, I_n controls the rate of increase of nutrients in the whole system, as can be seen by adding Equations (3.10a) and (3.10b) together after multiplying the latter by γ:

$$d(N + \gamma X)/dt = I_n - r_n N - \gamma e_1 X \tag{3.13a}$$

or, since N is held fixed at $N^* = k_1(d_1 + e_1)/(r_1 - e_1 - d_1)$

$$dX/dt = I_n - r_n N^* - \gamma e_1 X \tag{3.13b}$$

When $e_1 > 0$, X will approach an upper limit because the term $\gamma e_1 X$

increases until $dX/dt = 0$:

$$X^* = (I_n - r_n N^*)/(\gamma e_1)$$ (3.14, same as 3.9b)

When $e_1 = 0$, however, the right-hand side cannot be made to equal zero for any values of $I_n > r_n N^*$, since N^* is fixed and the loss rate of nutrient, $r_n N^*$, cannot balance the input. The total amount of nutrient in the system, $N + \gamma X$, must continually increase when both $e_1 = 0$ and $I_n > r_n N^*$. All of this increasing nutrient goes into autotroph biomass, so that $X^* \to \infty$ as $t \to \infty$.

Are the types of behaviour exhibited by the model (Equations 3.8) typical of any systems in nature? The strong positive relationship between nutrient input, I_n, and autotroph biomass, X^*, appears to be qualitatively supported by data on algal biomass in lakes. Welch et al. (1975) plotted algal biomass against input of the limiting nutrient on a log–log scale, phosphorus (g loading per unit volume) for a set of lakes in the United States and Sweden (Figure 3.5). Actually, the biomasses plotted in this graph were normalized by being divided by phosphorus residence time in the lake. The rationale for this normalization is that it helps correct for the fact that in lakes with small residence times, notably smaller lakes with a large throughput of water, the limiting nutrient has less chance to be taken up through the autotroph and through the food chain.

It is interesting to compare this empirical behaviour with the expression for standing stock autotroph biomass, X^*, from the model. This expression,

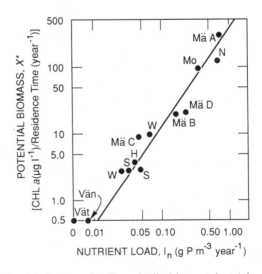

Figure 3.5 Relationship between loading of a limiting nutrient (phosphorus) per unit volume of lake water and biomass (expressed as CHL *a* or chlorophyll *a*, the most common type of chlorophyll, (normalized by nutrient residence time in the lake). (From Welch et al., 1975.)

shown in Equation (3.9b), predicts that X^* will increase linearly with nutrient loading, I_n. Recall that the approximate nutrient residence time for nutrient in the lake is $T_{res} = 1/r_n$. It is 'approximate' because it is the residence time that would exist in the absence of autotrophic biomass in the lake. According to the expression (3.9b) for X^*, the smaller the residence time is, or the larger r_n is, the smaller X^* will be. This agrees entirely with the argument of Welch et al. (1975) that a smaller residence time of phosphorus, and thus a smaller potential probability of uptake by autotrophs, will lead to smaller standing stocks of biomass. By multiplying the expression (3.9b) for X^* by r_n, which is equivalent to dividing it by the phosphorus residence time, one could possibly neutralize some of the effects of the $-r_n N^*$ term and make X^* relatively independent of different lake residence times. It is clear from the expression that this normalization will not do this precisely, but the effect would certainly be in the same direction as the results plotted by Welch et al. Therefore, we can say tentatively that the model prediction of a linear increase of autotrophic biomass as a function of per unit volume input of limiting nutrient, corrected for variations in residence time of the nutrient atoms in the system, reflects data on lake ecosystems.

One final point regarding Figure 3.5 is worth noting. This is that the increase in autotroph biomass with increasing nutrient input depends on a shift in the phytoplankton community from species that are predominantly edible to species that are predominantly non-edible to zooplankton with increasing nutrient. Otherwise, much of the increase in autotroph productivity occurring with increases in nutrient input would go straight into increases in herbivore production and not be seen in increases in autotroph biomass.

3.5   PRODUCTION RESULTING FROM NEW AND RECYCLED NUTRIENTS

An important distinction can be made between the amount of autotrophic production that utilizes nutrient molecules that have just entered the system from outside and the amount of production that uses nutrient molecules that have already circulated internally through the autotroph and detrital biomass at least once. The ratio of these quantities would give a measure of the dependence of production in the system on the recycling of nutrients. This ratio is also interesting to know because it can have some relationship to other properties of the system, such as stability and resilience.

The rates of inflow to the available nutrient pool of new and recycled nutrient molecules, respectively, are I_n and $\gamma d_D D^*$ (Figure 3.2) where, from Equation (3.9b,c)

$$\gamma d_D D^* = d_1(d_D + e_D)(I_n - r_n N^*)/[(d_1 e_D + d_D e_1 + e_1 e_D)] \qquad (3.15)$$

As the rates of loss of biomass from the system, e_1 and e_D, decrease, $\gamma d_D D^*$ increases, and recycling plays a greater role in production.

Finn (1976, 1978, 1980) has developed formal expressions, which he called the cycling efficiency and the cycling index, to describe differences in the cycling of different nutrient atoms in different ecosystems. These indices can be discussed in terms of our analysis. To do this it is useful to generalize the fluxes in Figure 3.1 by representing them as F_{ij}'s (in $g\,s^{-1}$), where i is the donor and j is a receptor compartment (Figure 3.6). Steady state will be assumed here, although Finn's approach is general enough to apply to non-steady-state systems also.

Finn (1976) defined cycling efficiency, RE_k, of a compartment k as the fraction of throughflow through the compartment that is recycled. To apply this to the model in Figure 3.6, note first that the throughflows for the nutrient pool, autotroph and detrital compartments in steady state are, respectively, $F_{ON} + F_{DN}$ (or equivalently, $F_{NO} + F_{NX}$), F_{NX}, and F_{XD}. Then the cycling efficiency of the nutrient pool compartment, or the ratio of recycled throughput (that which has passed through compartment N in the past and is reentering it) to total throughput (all nutrient entering the N compartment), is

$$RE_N = F_{DN}/(F_{ON} + F_{DN})$$

and, likewise, the cycling efficiencies for the other compartments are,

$$RE_X = (F_{XD}/F_{NX})(F_{DN}/F_{XD})[F_{NX}/(F_{ON} + F_{DN})]$$
$$= F_{DN}/(F_{ON} + F_{DN})$$
$$RE_D = (F_{DN}/F_{XD})(F_{XD}/F_{NX})[F_{NX}/(F_{ON} + F_{DN})]$$
$$= F_{DN}/(F_{ON} + F_{DN})$$

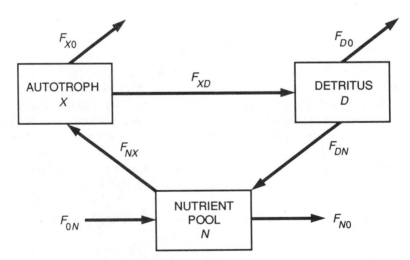

Figure 3.6 Schematic diagram of system with nutrient pool, autotroph and detritus compartments and general fluxes of nutrients between them.

To understand how RE_X and RE_D are derived, consider RE_X. The ratio F_{XD}/F_{NX} is the ratio of nutrient throughput of the autotroph that passes on to the detrital compartment, while F_{DN}/F_{XD} is the ratio of the nutrient throughput of the detrital compartment that passes to the nutrient pool, and $F_{NX}/(F_{ON} + F_{DN})$ is the ratio of the nutrient throughput of the nutrient pool that passes to the autotroph. The product of all three ratios, each of which is less than or equal to unity, is the fraction of nutrient molecules passing through the autotroph that returns to the autotroph. Similar logic holds for RE_D.

In this simple system, because all three compartments are in a single loop, the cycling efficiencies of all the compartments are identical. This has a further simplification regarding Finn's other index, the cycling index of the whole system, CI. Finn defines CI as

$$CI = TST_c/TST \tag{3.16}$$

where TST is the sum of all the throughputs of all the compartments in the system and TST_c is the sum of the cycling efficiencies of each compartment weighted by the throughput for that compartment. Thus, for the system in Figure 3.6

$$TST = F_{ON} + F_{DN} + F_{NX} + F_{DN}$$

Because the cycling efficiency for each compartment in Figure 3.6 is the same, TST_c is simply the sum of the same cycling efficiency, $F_{DN}/(F_{ON} + F_{DN})$, multiplied by each throughput;

$$TST_c = [F_{DN}/(F_{ON} + F_{DN})](F_{ON} + F_{DN} + F_{NX} + F_{DN}) \tag{3.17}$$

and, thus,

$$CI = F_{DN}/(F_{ON} + F_{DN}) \tag{3.18}$$

which, in this case, is the same as the cycling efficiency of each compartment.

The cycling index of nutrients in a system may be of less interest, for some purposes, than the mean length of time a nutrient molecule stays in the system; that is, the time between its first entry into the available nutrient pool N and its exit from the system through N, X or D. As described in Chapter 2, the mean residence time a nutrient molecule spends in a particular compartment per pass through that compartment is the standing stock of that compartment divided by the steady-state flux through the compartment. Analogously, for a multicompartment system, the mean residence time or turnover time of a nutrient molecule in the system in steady state is

$$T_{res} = N_T^*/I_n \tag{3.19}$$

where N_T^* is the total nutrient stored at any time in the steady-state system:

$$N_T^* = N^* + \gamma X^* + \gamma D^* \tag{3.20}$$

where N^*, X^* and D^* are given by Equations (3.9a,b,c).

It is interesting to look in the high recycling case to study how recycling affects the dynamics of the system. In the limit that I_n is very large and e_1 and e_D are very small, so that recycling is high (there is little loss of nutrient from the system due to biomass losses), the nutrient stored in biomass, $\gamma X^* + \gamma D^*$, far outweighs the available nutrient pool, N^*, so that N_T^* approaches

$$N_T^* = I_n[(d_D + e_D) + d_1]/(d_1 e_1 + d_D e_1 + e_1 e_D) \tag{3.21}$$

and

$$T_{res} = (d_1 + d_D + e_D)/(d_1 e_1 + d_D e_1 + e_1 e_D) \tag{3.22}$$

What this means is that T_{res} varies inversely with the loss rates e_1 and e_D and is independent of the nutrient input I_n and of the loss rate r_n from the nutrient pool. The reason for the independence from I_n is that an increase in I_n causes a proportional increase in N_T^*; thus turnover time will not change. The independence of T_{res} from r_n reflects the fact that the nutrient pool N^* is very small so that the nutrient loss rate $- r_n N^*$ is small and does not contribute to the residence time.

Later in this chapter, we show that T_{res} is closely related to the time it takes the system to return to equilibrium, T_R, following a perturbation. What is interesting is that T_{res} is independent of I_n, depending only on loss rates. For small values of I_n, it is easy to show that T_{res} can depend on I_n, although the precise form of this dependence is too complex to be predicted intuitively.

3.6 LOCAL STABILITY

The steady-state equilibrium point (N^*, X^*, D^*) was found above and some of its properties were explored. An additional feature of crucial importance is the local stability of this equilibrium point. As explained in Chapter 2, local stability refers to the tendency of system variables to return to the steady-state point following a very small perturbation.

It is possible to show that the set of Equations (3.8a,b,c) are locally stable. Unlike the model equations analysed in Chapters 1 and 2, Equations (3.8a,b,c) are non-linear and cannot be solved by the linear techniques used in Chapter 2. However, the equations can first be linearized about the steady-state equilibrium and the resultant linear equations can be solved and examined for stability.

To linearize Equations (3.8a,b,c), we introduce new variables, N', X' and D', which correspond to small deviations from equilibrium;

$$N' = N - N^* \quad \text{or} \quad N = N^* + N' \quad (N' \ll N^*) \tag{3.22a}$$

$$X' = X - X^* \quad \text{or} \quad X = X^* + N' \quad (X' \ll X^*) \tag{3.22b}$$

$$D' = D - D^* \quad \text{or} \quad D = D^* + D' \quad (D' \ll D^*). \tag{3.22c}$$

When N, X and D are substituted for in terms of N', X' and D' in Equations

(3.8a,b,c), the resultant equations are (see Appendix C)

$$dN'/dt = - r_n N' - \gamma r_1 k_1 X^* N'/(k_1 + N^*)^2 - \gamma r_1 N^* X'/(k_1 + N^*)$$
$$+ \gamma d_D D' \tag{3.23a}$$

$$dX'/dt = r_1 k_1 X^* N'/(k_1 + N^*)^2 \tag{3.23b}$$

$$dD'/dt = d_1 X' - (d_D + e_D)D' \tag{3.23c}$$

The equation for the eigenvalues of this set of equations is given by the determinant

$$\det \begin{vmatrix} -r_n - \dfrac{\gamma r_1 k_1 X^*}{(k_1 + N^*)^2} - \lambda & -\dfrac{\gamma r_1 N^*}{k_1 + N^*} & \gamma d_D \\[2ex] \dfrac{r_1 k_1 X^*}{(k_1 + N^*)^2} & -\lambda & 0 \\[2ex] 0 & d_1 & -(d_2 + e_2) - \lambda \end{vmatrix} = 0 \tag{3.24}$$

It can be shown (Appendix D) that this eigenvalue equation only has roots λ that have negative real parts, so that the system is always stable.

The local stability of pairs of autonomous equations such as Equations (3.10a,b) (autonomous meaning that none of the coefficients are time-dependent) can often be determined graphically by noting the slopes of the intersecting zero isoclines. This will be described in Chapter 5.

The fact that Equations (3.8a,b,c) are stable does not mean that the dynamic behaviour of this system is uninteresting. In fact, Equations (3.8a,b,c) can produce striking behaviour under typical natural conditions. O'Brien (1974) applied a model similar to this one to a single phytoplankton population starting at low numbers in a nutrient-rich medium, such as might occur in a pond or lake in spring.

O'Brien's (1974) model differs from Equations (3.8a,b,c) in several ways: (1) the detrital component is ignored; (2) nutrient feedback from organic matter to the nutrient pool is ignored; (3) the loss of nutrient from the lake by outflow, $r_n N$, is ignored; and (4) instead of autotroph biomass, cell number is chosen as the variable, so that r_1 (g_m in O'Brien's model) represents the rate of increase in cell numbers (assumed proportional to total biomass) rather than biomass growth directly. Thus, the parameters r_n, d_1, d_D and e_D are set to zero in Equations (3.8a,b,c). The remaining parameter values are, in current notation,

$r_1 = 0.5$ cell divisions d^{-1}

$d_1 = 0.15$ d^{-1}

$k_1 = 10$

$I_n = 2\,\mu g\,l^{-1}\,d^{-1}$

$\gamma = 2.36 \times 10^{-7}\,\mu g\ cell^{-1}$

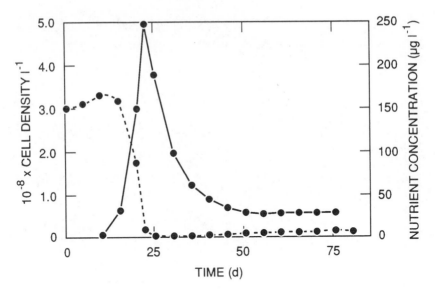

Figure 3.7 Cell number density as numbers l^{-1} (solid line) and nutrient concentration as $\mu g\,l^{-1}$ (dashed line) through time from simulations of Equations (3.8a,b) with parameters similar to those used in Figure 1 of O'Brien (1974).

Initial values of nutrient concentration $N = 150\,\mu g\,l^{-1}$ and $X = 500\,000$ cells were chosen. The results of the simulation over 100 days are shown in Figure 3.7. The phytoplankton population, starting from very low numbers, exploded in size within several weeks to over 5×10^8 cells per litre, causing a drastic decline in the available nutrients. This led to a crash of the phytoplankton population towards a steady-state equilibrium level, accompanied by a gradual rise in nutrient concentration towards steady state. As O'Brien (1974) pointed out, the model results indicate that phytoplankton crashes can easily be caused through nutrient limitation. The presence of zooplankton consumers is not a necessary factor.

The rate at which the system eventually settled to steady state is its resilience. Some generalizations concerning system resilience are discussed in the next section.

3.7 RESILIENCE

If a model system is shown to be locally stable, as is the case for Equations (3.8a,b,c), the question of next greatest significance is how resilient the system is, that is, how fast it returns to the steady state following a perturbation (or the inverse of the return time, T_R; see Chapter 2). Resilience has received attention from theoretical ecologists because of its possible role in the structure of ecological food webs.

Imagine a region of land receiving abundant rainfall and solar radiation but only a very low rate of input of essential nutrients from the outside (in solution in precipitation or from rock weathering). Suppose that when the system reaches steady state, trees growing on this land have extremely efficient mechanism of recovery and retention of nutrients that are released in the soil through litter decomposition, so that only a tiny percentage of released nutrient is lost through leaching and runoff. This efficient recycling of nutrient enables a majestic forest of massive trees to grow, although it may take thousands of years to reach steady state because of the small input of nutrients. If this steady-state forest is perturbed by removal of, say, one-half of the biomass from the site, as long as the parameters of uptake and loss remain unaltered (which may not be the case in real situations), it will take thousands of years to replace the lost nutrients. In particular, the turnover time, $T_{res} = N_T^*/I_n$, where N_T^* is the total amount of nutrient stored in the biomass, litter and soil, while I_n is the input of the limiting nutrient, is a good estimate of the return time to steady state, as will be shown below.

If, under the same rate of nutrient input by precipitation and rock weathering, instead of a high steady-state rate of nutrient recovery and recycling, the forest system tends to lose nutrients rather rapidly and not to recycle them more than a few times on average, the steady-state forest biomass will be relatively small. Now the turnover time, $T_{res} = N_T^*/I_n$, will be proportionally lower than in the first case. If one-half of the biomass of the trees is removed, it should take a much shorter time for this amount of biomass to recover.

These two examples illustrate the concept of resilience, or the rate of recovery, as a function of the turnover time of a limiting nutrient. Resilience is often roughly the inverse of turnover time, or $1/T_{res}$.

Some results concerning the effect of nutrient limitation and recycling on food web resilience have already been derived from models (Jordan et al., 1972; Dudzik et al., 1975; DeAngelis, 1980; Harwell et al., 1981). It was shown that the return time of the system tends to be negatively related to the rate of nutrient input, I_n, to the system per unit standing stock of system biomass (to which the standing stock of nutrients, N_T^*, is proportional). This is an intuitively reasonable conclusion since a faster rate of nutrient input per unit biomass should decrease the nutrient turnover time and thus increase the rate at which a system can recover from a loss of biomass.

To examine the resilience of the system in Figure 3.2, Equations (3.8a,b,c) were solved numerically on a computer. The set of parameters shown in Table 3.3 was used for all of the simulations and analyses performed on the model equations. These parameters are hypothetical and no units are assumed. However, the values, viewed as a whole, form a plausible set for describing typical dynamic behaviour of a nutrient-limited system. We describe here the observed resilience characteristics of the perturbed system as a function of I_n.

The return time T_R of the system after a removal of 10% of each compartment from the steady state was computed for a series of values of

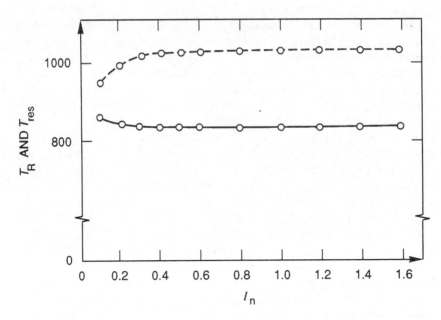

Figure 3.8 Return time, T_R (solid line) and nutrient turnover or residence time, $T_{res} = N_T^*/I_n$ (dashed line) for system described by Equations (3.8a,b,c), as a function of nutrient input, I_n ($\times 10^3$). Parameter values are specified in Table 3.3.

I_n by solving Equations (3.8) numerically and then using expression (2.8). The return time as a function of I_n is shown in Figure 3.8 (solid line). The value of T_R is observed to remain at a relatively constant level for all values of I_n that are large enough for the autotroph to persist.

To attempt to interpret the behaviour of T_R as a function of I_n, let us see if the generalization $T_R \cong T_{res} = N_T^*/I_n$, discussed earlier (Equation 3.19) has any explanatory power here. In this case, $N_T^* = N^* + \gamma X^* + \gamma D^*$, where N^*, X^* and D^* are the steady-state values. We have already solved for these values [see Equations (3.9)]. Note that the available nutrient is held at a fixed level by the autotroph, so that it does not increase as I_n increases (Figure 3.3). On the other hand, X^* and D^* are increasing functions of I_n.

When N_T^*/I_n is plotted as a function of I_n, this ratio is roughly constant as a function of I_n for all values of I_n (dashed line in Figure 3.8). For large values of I_n, the closeness of the return time predicted by N_T^*/I_n (dashed line) to T_R calculated from Equation (2.8) is remarkable, since, in a non-linear system, N_T^*/I_n is expected to provide only a rough estimate of T_R.

The resilience of forest ecosystem with respect to nutrient loss must be considered in commercial harvesting of trees. For example, whole-tree harvesting of southern New England hardwoods may remove 530 kg ha^{-1} of calcium directly, plus additional runoff losses of about 28 kg ha^{-1} (Tritton *et al.*, 1987). Only 2 kg ha^{-1} year^{-1} of calcium are added by precipitation. A much larger amount is probably added by weathering of soils and bedrock, but since this is not known accurately, it cannot be excluded that the recovery time of the forest following whole-tree harvesting will be slow.

3.8 FURTHER CONSIDERATIONS

The Monod-type function used for $r(N)$ in the above model [Equation (3.7)] has the virtue of simplicity, but that does not mean that it provides the only possible description of uptake. Alternative models of uptake have been examined and found useful.

One obvious alternative follows from noting that growth of the autotroph may be impossible to sustain unless the level of limiting nutrient exceeds some critical level N_c. In this case, $r(N)$ has a negative value until N exceeds N_c. The function

$$r(N) = r_1(N - N_c)/[k_1 + (N - N_c)] \tag{3.25}$$

can be used to represent this situation. When this function is used in Equations (3.10), which are a truncated form of Equations (3.8) in which the detritus compartment is ignored, the zero isoclines that are equivalent to Equations (3.11a,b) are

$$N = N_c + k_1(e_1 + d_1)/(r_1 - e_1 - d_1) \tag{3.26a}$$

$$X = (I_n - r_n N)[k_1 + (N - N_c)]/[r_1(N - N_c) - d_1] \tag{3.26b}$$

While these zero isoclines differ quantitatively from Equations (3.11a,b), the basic shape of the curves is the same (Figure 3.4) and the properties of stability and resilience are similar to those discussed for Equations (3.10a,b).

3.9 SUMMARY AND CONCLUSIONS

The introduction of an autotrophic component complicates the dynamics of material elements that are nutrients for the autotroph. Because autotroph biomass actively accumulates the nutrients from the available pool and grows in size, the equations describing the system are non-linear. Nutrient limitation of autotrophic growth is common in natural systems, with nitrogen or phosphorus commonly being the limiting bioelement. Many types of adaptations have evolved in autotrophs to alleviate this limitation on growth due to the scarcity of one or more nutrients, including symbiotic associations with mycorrhizal fungi. The effect of most of these adaptations is to increase the proportion of biomass production in the autotroph that utilizes recycled nutrients, either through cycling within the autotroph itself or in biogeochemical cycles between the soil and aboveground and below-ground living matter.

A model of the dynamics of a nutrient–autotroph–detritus system was constructed in which nutrient-limited autotroph growth was described by a Monod growth function and which included recycling of nutrients from detritus back to the nutrient pool. The model was analysed for steady-state solutions, local stability and resilience. One of the important properties of the steady-state solution was that the available nutrient level stays constant as a function of nutrient loading, but the autotroph biomass increases linearly. This prediction with regard to autotroph biomass seems to be corroborated qualitatively by data from a number of lakes.

4 Nutrients and autotrophs: variable internal nutrient levels

4.1 INTRODUCTION

The mathematical formulations of nutrient-limited autotrophic growth used in Chapter 3, the Monod-type growth function and its variations, are based on the assumption that the growth rate depends directly on the pool of available limiting nutrient in solution external to the autotroph. This assumption is not always a good one, at least on short time scales, over which growth may depend directly on the level of nutrient stored internally in the autotroph. Another assumption of Chapter 3 was that a nutrient occurs as a fixed proportion of autotroph biomass. While this assumption makes model analysis easier, it ignores such facts as the widely different nutrient requirements of different species, the different nutrient storage levels between different individuals of the same species in different environments, and the allocations of nutrients to different components of the same plant.

The present chapter describes modelling approaches that allow for different internal nutrient storages. The result of this generalization is that new types of dynamic phenomena are encountered. These include the nutrient-related processes in the same model system taking place on vastly different time scales and the occurrence of structural instabilities or catastrophes.

4.2 VARIABILITY IN PLANT RESPONSES TO NUTRIENTS

Careful studies on phytoplankton in chemostats indicate that the specific growth rate depends directly on the level of internal nutrient, or that in solution within the autotroph cells, rather than on the external level (Collins, 1980). In particular, the need for explicitly taking into account the internal nutrient level was demonstrated with the limiting nutrient in external solution in chemostat experiments was measured to be much less than expected when the Monod growth model was applied to growth measurements of algae (McCarthy and Altabet, 1984). Rhee (1973) showed, in fact, that there could be an exponential phase of algal growth even after phosphate was exhausted from the surrounding medium.

The explanation for these observations, as Ketchum (1939) discerned, is that plants are able to engage in 'luxury consumption', that is, they can store nutrients above their immediate requirements that are then used when external nutrient levels decline. For example, phosphate ions, PO^{4-}, can be stored in cells at levels much greater than needed for immediate metabolic demands (Welch, 1980), and the maximum uptake rate of some nutrients is 2 to 20 times larger than the maximum growth rate of the autotroph that uses them (Droop, 1968; Eppley and Thomas, 1969; Caperon and Meyer, 1972a,b).

Storage and internal recycling are also obvious adaptive responses to nutrient limitation problems of large perennial plants. When another factor, such as light, is limiting growth, nutrients may be taken up by plants in excess of immediate metabolic needs as luxury consumption and stored in high concentrations in the foliage (e.g. Waring and Schlesinger, 1985).

Tundra plants are frequently adapted to conditions of low nutrient availability. Fertilization can lead to considerable levels of accumulation in such tundra species as cotton grass (Goodman and Perkins, 1959). Numerous studies have shown that the addition of fertilizer to forest ecosystems increases the nutrient concentrations in green foliage of trees, although this varies to some extent with species (e.g. Khanna and Ulrich, 1981). Studies have indicated that nutrients such as N and P vary through the growing season in plants, being taken up in luxury consumption when they are most available in the soil (e.g. Cole *et al.*, 1978; Howard-Williams and Allanson, 1981; Hopkinson and Shubauer, 1984).

Evergreen leaves appear to act as reservoirs for nutrients storage both seasonally and from year to year (Chapin *et al.*, 1980; Gray, 1983), which is an adaptation to nutrient deficiency. Gray and Schlesinger (1983) contrasted the annual nutrient regime of the evergreen shrub *Ceanothus megacarpus* Nutt. in the California chaparral with the drought-deciduous shrub *Salvia leucophylla* of the mountains of southern California. The deciduous plant is closely coupled to nutrient availability in the soil and is able to take advantage of nitrogen flushes during certain times of the year. *Ceanothus*, which grows on soils low in nutrient, shows little growth response to experimental variation in nitrogen level. Its high internal storage of nutrients (largely in foliage) and its low loss rates of leaching from leaves allows *Ceanothus* to be relatively decoupled from variations in soil nutrient availability.

Early successional species may show higher production as well as higher levels of nutrient concentration in foliage than later successional species, as Turner *et al.* (1976) showed in comparisons of red alder (*Alnus rubra* Bong.) and Douglas-fir (*Pseudotsuga menziessii*) in the northwestern US (Washington state).

Clarkson (1967) demonstrated differences in nutrient accumulation of two grass species, *Agrostis setacea* and *Agrostis stolinifera*, grown in a series of phosphorus supply levels. In *A. setacea*, which is a native of infertile acidic grasslands, phosphorus accumulated within the plant as it increased in the external medium. On the other hand, *A. stolonifera* did not accumulate phosphorus, but converted it directly into growth (Figure 4.1).

Different environmental conditions frequently lead to different nutrient allocations to organic matter within plants. Black and Wight (1979) applied N and P fertilizer to a mixed grass and shrub community in the northern Great Plains over an eight-year period. Although the yield of total biomass levelled off before the end of the period, continually higher levels of protein were produced in the fertilized plots throughout the entire period. Tanner (1985), studying Jamaican montane forests on two sites (Mor Ridge forest and Mull Ridge forest) similar in elevation and rainfall, found great differences in structure between the sites, with the Mor Ridge forest impoverished in size and nutrient concentrations in leaves and stems, especially with respect to N and K. The concentrations of these nutrients found in the Mor Ridge forest were less than two-thirds the concentrations in the Mull Ridge forest.

Thus, there is abundant evidence that the internal concentrations of potentially limiting nutrients vary significantly in autotrophic organisms of many taxa. The implication is that internal nutrient levels must be considered as variables in models if nutrient uptake and growth by autotrophs is to be properly described. Several ways of doing this have been proposed. Here we will consider a few of these and examine their consequences to the system dynamics of food chains abbreviated to include only the autotroph level.

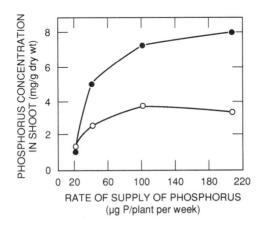

Figure 4.1   Phosphorus concentrations in shoots as a function of rate of supply for two different grass species, *Agrostis setacea* (closed circles) and *Agrostis stolinifera* (open circles). (Adapted from a table by Clarkson, 1967.)

4.3   FORMULATION OF MODEL

A large number of authors have modelled internal nutrient storage. These include Droop (1968), Fuhs (1969), Caperon and Meyer (1972a,b), Rhee (1973), Lehman *et al.* (1975), DiToro (1976) and Bierman (1976). It would consume a great deal of time to compare all of these formulations. Instead, we will focus first on Droop's model.

Assume, as in Chapter 3, that N is the external amount of limiting nutrient in the available pool, and that X is the amount of autotroph biomass in the system. Growth will not be assumed to be a direct function of external nutrient level N. Instead, the process is considered to take place in two steps: (1) nutrient uptake by the cell and (2) growth that is a function of nutrient storage in the cell. Nutrient uptake has been measured in batch culture experiments by placing nutrient-depleted algae in a fresh medium with measured nutrient levels. The algae quickly take up the nutrients. The uptake rate as a function of nutrient concentration typically follows a curve such as that in Figure 4.2 (from Droop, 1973), which has the form of the simple hyperbolic function familiar from Michaelis–Menten kinetics. This uptake rate function is written as

$$V = V_{max}N/(k_1 + N) \tag{4.1}$$

where V is the uptake rate, V_{max} is the maximum uptake rate, and k_1 is the half-saturation constant. When V is the uptake rate for cells, it can be expressed as nutrient mass per cell per unit time or mass per unit biomass per unit time (e.g. $mg\,g^{-1}\,d^{-1}$). At high concentrations, the nutrient overwhelms the capacity of cell enzymes to transport nutrient ions into the cell.

The next step is to relate uptake to changes in the cell nutrient concentration, termed the 'cell quota', or amount of nutrient per unit biomass, Q (usually having units of $mg\,g^{-1}$ or $\mu g\,g^{-1}$) Define μ as the rate of change in grams of biomass per unit standing stock of biomass per unit time. Then it

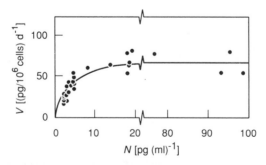

Figure 4.2   Specific rate of uptake, V, of vitamin B_{12} by *Monochrysis lutheri* as a function of substrate concentration, N. The nutrient level has units of pg ml^{-1} and the uptake rate is in pg/10^6 cells per day. [From Droop (1973).]

follows that the equation describing Q is

$$dQ/dt = V - \mu Q \qquad (4.2)$$

The second term in Equation (4.2) represents the decrease in concentration that accompanies growth in biomass. That is, as biomass increases and the amount of nutrient in the biomass stays the same, Q decreases in proportion. Droop found that the term μQ, when measured experimentally as a function of Q, is consistently linear and can be fitted to a regression equation

$$\mu Q = \mu_m (Q - k_Q) \qquad (4.3)$$

where μ_m and k_Q are constants. This equation can be rewritten as

$$\mu = \mu_m [1 - (k_Q/Q)] \qquad (4.4)$$

Data on μ versus Q for two different nutrients are shown in Figure 4.3. This equation demonstrates that a certain minimum cell nutrient quota, $Q = k_Q$ is necessary for growth to occur (i.e. for μ to be a positive value) and it demonstrates that increases in Q above that level cause μ to increase and eventually asymptote to μ_m.

These empirical relationships can now be integrated into the system of equations for the nutrient pool, N, the autotroph biomass, X, and the detritus, D, in analogy with what was done using the Monod growth function in Chapter 3. A few comments should be made before writing these equations. First, the uptake rate V refers to uptake per unit biomass. To represent the removal of nutrient from the nutrient pool correctly, V [Equation (4.1)] must be multiplied by autotroph biomass X. Second, μ [Equation (4.4)] represents the rate of change of biomass per unit biomass of the autotroph per unit time. This value must be multiplied by X to represent the rate of growth of autotroph biomass in absolute terms ($\mathrm{g\,d^{-1}}$,

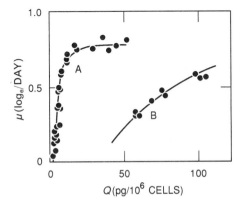

Figure 4.3   Specific growth rate, μ, of *Monochrysis* as a function of cell quota, Q, of vitamin B_{12}. A, Chemostat with vitamin B_{12} limiting. B, Chemostat with nitrate limiting. The units of Q are pg per 10^6 cells and the units of μ are $\mathrm{d^{-1}}$.

for example). Third, it is convenient for the present to let the nutrient concentration in the detritus stay at the same value, Q, as the nutrient concentration in the living autotroph biomass, at least for analysis of steady state properties of the model, (a detailed consideration of decomposition in Chapter 7 will modify this assumption). The rate of return of nutrient from the detritus to the nutrient pool is thus $d_D QD$. Fourth, the equation for the rate of change of nutrient concentration in biomass, Q, is given by Equation (4.2).

The set of equations equivalent to Equations: (3.8a–c) is now

$$\mathrm{d}N/\mathrm{d}t = I_n - r_n N - [V_{max} NX/(k_1 + N)] + d_D QD \tag{4.5a}$$

$$\mathrm{d}Q/\mathrm{d}t = [V_{max} N/(k_1 + N)] - \mu_m(1 - k_Q/Q)Q \tag{4.5b}$$

$$\mathrm{d}X/\mathrm{d}t = \mu_m(1 - k_Q/Q)X - (d_1 + e_1)X \tag{4.5c}$$

$$\mathrm{d}D/\mathrm{d}t = d_1 X - (d_D + e_D)D \tag{4.5d}$$

Every term in Equation (4.5b) has units of amount of nutrient per unit biomass per unit time (such as, $\mathrm{mg\,g^{-1}\,s^{-1}}$).

In Droop's (1968) model, therefore, cell growth is considered to be a two-step process, involving separate nutrient uptake and biomass synthesis mechanisms. There are alternative models in which the nutrient uptake function, or the relationship between cell nutrient concentration, Q, differs from Droop's assumptions, but the basic concept in these other models is the same.

The next section will explore the behaviour of this new model of a nutrient–autotroph–detritus system, in particular to see if it differs appreciably from the system incorporating the Monod growth function analysed in Chapter 3.

4.4 STEADY-STATE SOLUTIONS

By putting the right-hand sides of Equations (4.5) equal to zero, one can solve for the steady-state equilibrium values of the four variables. These are

$$Q^* = k_Q \mu_m/(\mu_m - d_1 - e_1) \tag{4.6a}$$

$$N^* = Q^* k_1(d_1 + e_1)/[V_{max} - Q^*(d_1 + e_1)] \tag{4.6b}$$

$$X^* = (I_n - r_n N^*)(d_D + e_D)/[Q^*(d_1 e_D + d_D e_1 + e_1 e_D)$$

$$= (I_n - r_n N^*)/(Q^* e_1) \quad (\text{if } e_D = 0) \tag{4.6c}$$

$$D^* = d_1 X^*/(d_D + e_D) \tag{4.6d}$$

By comparing Equations (4.6b,c,d) with the steady-state solutions of the Monod growth formulation in the preceding chapter (Equations 3.9a,b,c), one can see that the expressions for N^*, X^* and D^* are equivalent in the two cases if

$$Q^* = \gamma \tag{4.7a}$$

$$V_{\max} = \gamma r_1 \tag{4.7b}$$

Burmaster (1979) demonstrated in general the mathematical equivalence of the steady-state properties of three commonly used models: the two already mentioned (Monod's and Droop's models), as well as a model developed originally by Caperon (1967). Hence, each model is in Burmaster's words 'an equally valid (or equally invalid) formal description of algal growth at steady state. A corollary to this is that it is not possible to distinguish between internal and external nutrient pools as the determinant of growth from steady state measurements'.

It is interesting to note that, in particular, as in the Monod model, N^* in Equation (4.6b) is independent of I_n. It is controlled from top-down by the autotroph. Autotroph biomass X^*, on the other hand, increases linearly with increasing I_n.

The next sections will consider Equations (4.5) from the dynamic point of view, where interesting differences from the Monod growth model might occur.

4.5 LOCAL STABILITY

The inclusion of internal nutrient concentration as a variable did not change the fundamental nature of the steady-state solutions of the present model compared with the model based on the Monod growth function. It remains to be seen whether or not the local stability characteristics are altered in the new formulation that takes into account the internal nutrient pool.

We found in Chapter 3 that when the Monod growth function is used for autotrophic growth, the system is always stable. The same type of analysis can be used for Equations (4.5). Here I will 'cheat' and analyse a simpler version of the equations by ignoring the detrital compartment [Equation (4.5d)] and assuming that nutrient is recycled directly from the autotrophs to the available nutrient pool without passage through detritus. This requires that $d_1 QX$ replaces $d_D QD$ in Equation (4.5a). Reduction of the system from four to three variables greatly simplifies stability analysis. The stability can be determined by linearizing Equations (4.5a,b,c) about the equilibrium point (N^*, Q^*, X^*) and solving for the eigenvalues. It is left as an exercise for the reader to determine that the eigenvalue equation is

$$\det \begin{vmatrix} r_n - \dfrac{V_{\max} k_1 X^*}{(k_1 + N^*)^2} - \lambda & d_1 X^* & -\dfrac{V_{\max} N^*}{k_1 + N^*} + d_1 Q^* \\[3ex] \dfrac{V_{\max} k_1}{(k_1 + N^*)^2} & -\mu_m - \lambda & 0 \\[3ex] 0 & \dfrac{\mu_m k_Q X^*}{Q^2} & -\lambda \end{vmatrix} = 0 \tag{4.8}$$

or, after expansion of Equation (4.8) and use of Equation (4.6b),

$$\lambda^3 + \left(r_n + \frac{V_{max}k_1 X^*}{(k_1 + N^*)^2} + \mu_m\right)\lambda^2 + \left(r_n\mu_m + \frac{V_{max}k_1\mu_m X^*}{(k_1 + N^*)^2} - \frac{V_{max}d_1 k_1 X^*}{(k_1 + N^*)^2}\right)\lambda$$

$$+ e_1\frac{V_{max}k_1}{(k_1 + N^*)^2}\frac{\mu_m k_Q X^*}{Q} = 0 \tag{4.9}$$

From the equilibrium solution Equation (4.6a) it can be observed that μ_m must exceed $d_1 + e_1$ in order for the steady-state value Q^* to be positive. Therefore, the coefficient of λ in Equation (4.9) is positive, so that all coefficients in Equation (4.9) are positive. A theorem from matrix theory then implies that the real parts of all the eigenvalues of Equation (4.9) are negative (e.g. Murata, 1977, p. 89), so that the system is stable for all parameter values for which the steady-state solutions are positive, that is, for all non-trivial solutions.

One naturally would like to know whether or not simplifying the model by omitting the detritus compartment made any difference in stability. I examined the problem with detritus included and obtained a fourth-order equation for λ analogous to Equation (4.9), but could not find a simple way to prove that all of the real parts of the eigenvalues were negative. Thus I am forced to acknowledge the possibility that the additional variable for detritus, even though it is involved only as linear donor-dependent terms in the equations, may make the system susceptible to unstable oscillations. Additions of compartments that lengthen the number of linkages in the cycling of nutrients from the autotroph back to the available nutrient pool are likely to increase the probability of instability.

Despite the stability of Equations (4.5), this model is capable of dynamic behaviour that is as interesting as that of Equations (3.8), with the added twist that the cell nutrient concentration is now a variable. It is useful to examine this behaviour, even though the Droop model is probably best used only for the examination of steady-state behaviour.

Parameter values were assigned to Equations (4.5) to make this model based on Droop's model similar to O'Brien's (1974) model, discussed in Chapter 3 (see Figure 3.7 and the parameters discussed nearby in the text), based on Monod's growth function. Thus, the values $I_n = 2.0\,\mu g\,l^{-1}\,d^{-1}$, $k_1 = 10$, $\mu_m = 0.5\,d^{-1}$, and $d_1 = 0.15\,d^{-1}$ were chosen, while $d_D = 0$, $e_1 = 0$, and $e_D = 0$. The variable X represents the number of cells rather than biomass. The minimum cell quota coefficient, k_Q, was set to 1.6×10^{-7}, so that the nutrient level in a cell at steady-state would be $2.36 \times 10^{-7}\,\mu g$, as assumed in the simulation of Equations (3.8) shown in Figure 3.7. The maximum uptake rate, V_{max}, could be assigned somewhat arbitrarily. As long as this uptake rate is much faster than the maximum biomass growth rate, μ_m, it should have no major effect on the growth of biomass. The rate V_{max} was assigned the value of $V_{max} = 0.000004\,\mu g\,d^{-1}$, which assumes that the nutrient in solution inside the hypothetical cell could turn over about 20 times in a day. Thus the time scale for appreciable growth of the cell quota can be assumed to be the order of an hour.

The results of the model simulation over a fifty-day period are shown in Figure 4.4a,b. Note that the dynamics occur on two distinct time scales. The first is a short time scale of a few days, during which the cell quota increases from a low initial level to a high level. Over this time period the external nutrient level, N, declines slightly, but the number of cells, X, is virtually unchanged. The dynamics of X are controlled, by μ_m, which means that appreciable growth occurs only on a time scale of weeks. Once X starts

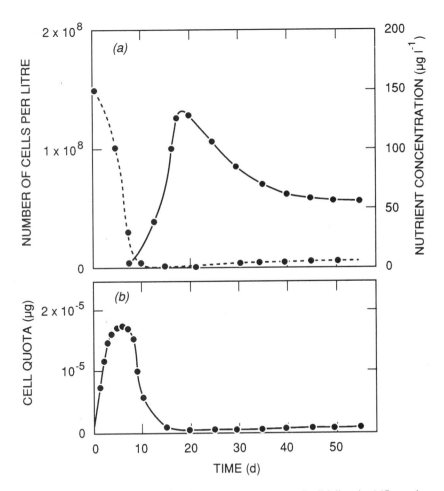

Figure 4.4 Simulation of a model of cell numbers, X, [solid line in (a)] nutrient concentration, N, [dotted line in (a)] and cell quota, Q, based on Droop's (1973) model [Equations (4.5)]. Parameter values were chosen to be similar to the simulations of the Monod-type growth model shown in Figure 3.7.

growing rapidly, it dominates the dynamics of N and Q. The external nutrient level plunges rapidly as more cells are created that can take up nutrient. The cell quota, Q, also decreases rapidly, as nutrient input I_n is unable to compensate for the rapid increase in biomass. After a couple of weeks, the cell quota has declined enough to start to limit biomass growth and X crashes. As X declines to its steady-state value, Q and N increase to theirs on the same time scale.

The dynamics of Droop's model shown in Figure 4.4 are reasonably similar to the simulation of the Monod-based model in Figure 3.7. The addition of the internal nutrient concentration appears to have slightly suppressed the biomass peak.

The resilience characteristics of this model are investigated in more detail in the next section.

4.6 RESILIENCE

The resilience of the system of autotroph, nutrient pool and detritus is a property that could be quite different in the cell quota model than in the Monod model for growth studied in the preceding chapter. This is because, while in the Monod growth model the autotrophic growth rate responded immediately to changes in the nutrient pool, in the cell quota model the step between nutrient uptake and growth introduces a lag.

Intuitively, one might expect that smaller values of μ_m and d_1 would decrease the resilience of the system, because this would slow the response of autotrophic growth to an increase in the cell nutrient quota Q [see

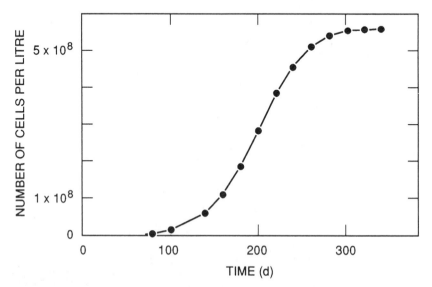

Figure 4.5 Values of cell numbers for a simulation of Equations (4.5). The parameters μ_m and d_1 were set to one-tenth the values in Figure 4.4.

Equation (4.5b)]. This idea is easily tested by performing simulations of the model using changed values of the two parameters. In particular, $\mu_m = 0.05$ and $d_1 = 0.015$ were used in Equations (4.5a,b). Figure 4.5 shows the simulation of X for these values. Not only does the system take longer to reach equilibrium, but the biomass, X, no longer exhibits a boom and crash behaviour but approaches a large steady-state value monotonically. The reason for this relatively placid behaviour is that the biomass growth rate is now slow enough not to drastically overshoot the capacity of the nutrient input.

Changes in the driving function of nutrient input, I_n, can also be made and the effects on resilience observed. As in the simulations on the Monod-type growth model in Chapter 3, return time can be shown to be virtually constant with respect to I_n. Biomass, X, however, is an increasing function of I_n.

4.7   THE POSSIBILITY OF CATASTROPHIC BEHAVIOUR

What is a 'catastrophe'? In the mathematical sense, a catastrophe is a kind of structural instability in a system. A system is structurally unstable when small changes in a driving variable or parameter can cause a sudden instability of the system. Structural instabilities can occur in models that have a certain degree of complexity.

The model represented by Equations (4.5), though it is a step in complexity beyond the Monod growth model because it explicitly includes the internal nutrient pool, is still a great simplification of the way in which the nutrient pools inside and outside the plant affect its growth. One of the simplifications is the assumption that increasing internal nutrient level always increases the growth rate of the autotroph. In fact, increasingly high values of almost any nutrient will eventually begin to have a toxic effect on the autotroph. Taking this into account, as Gatto and Rinaldi (1987) have done, leads to complex system behaviour, including the possibility, in some systems, of a fold catastrophe (*sensu* Zeeman, 1972), as will be shown below.

To illustrate the type of dynamics that can occur, I will outline part of Gatto and Rinaldi's (1987) model of the possible effects of nutrient enrichment on forest dynamics. Gatto and Rinaldi's model was developed for a generalized forest ecosystem, but is applicable to autotrophs in general. The model omits the detrital compartment, allowing recycling from the autotroph directly to the available nutrient pool. Gatto and Rinaldi's equations, modified to be consistent with the notation of this book, are:

$$dN/dt = I_n - r_n N - r_1 NX + d_1 XQ \tag{4.10a}$$

$$dQ/dt = r_1 N - f(Q)h(X)Q \tag{4.10b}$$

$$dX/dt = f(Q)h(X)X - (d_1 + e_1)X \tag{4.10c}$$

where $h(x)$ is the maximum production rate of new biomass per unit biomass and $f(Q)$ is the efficiency of the production, as affected by the cell nutrient concentration. Note that the Lotka–Volterra term, $r_1 NX$, in Equation (4.10a) has replaced the Michaelis–Menten term of Equations (4.5a). Aside from the Lotka–Volterra term in Equation (4.5b), Equations (4.10b,c) would be identical with Equations (4.5b,c) if $f(Q)$ were equal to $\mu_m(1 - k_Q/Q)$ and $h(X)$ were equal to 1.

Gatto and Rinaldi merely specify $f(Q)$ as an increasing and saturating function of Q. Equations (4.10a,b,c) are thus a more general case than Equations (4.5a,b,c). The authors assume that $h(X)$ is a decreasing function of X, based on data showing that the maximum production of tree biomass per unit biomass declines as a function of tree biomass (Figure 4.6).

Gatto and Rinaldi (1985) further assumed that mortality, $d_1 + e_1$, could be a function of either the external nutrient level N (which could damage roots) or the internal nutrient level Q (which in excessive levels could have a toxic effect). Only the latter effect will be considered here, so that the mortality rate is represented by allowing mortality rates, d_1 and e_1, to be increasing functions of Q, $d_1(Q) + e_1(Q)$. These increase faster than linearly for large enough values of Q.

The nature of the steady-state equilibrium of this system can be found by plotting zero isoclines on the Q,X-plane. To do this, first use the steady-state solution of Equations (4.10b,c) to express X in terms of Q. By combining Equations (4.10b,c), one can solve for N in terms of

$$N = [d_1(Q) + e_1(Q)]Q/r_1. \tag{4.11}$$

The right-hand side of Equation (4.10c) yields,

$$h(X) = [d_1(Q) + e_1(Q)]/f(Q), \tag{4.12}$$

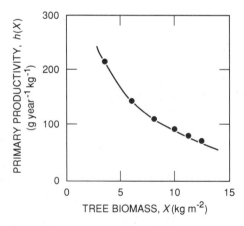

Figure 4.6 Data showing the decrease in primary production per unit biomass, $h(X)$, as a function of tree biomass, X. (From Gatto and Rinaldi, 1987.)

or, by solving for X, by taking the inverse of function h,

$$X = h^{-1}[(d_1(Q) + e_1(Q))/f(Q)] \tag{4.13}$$

One can make a qualitative guess as to how Equation (4.13) might look. First, recall the assumption that $f(Q)$ increases and saturates as Q increases, while $d_1(Q) + e_1(Q)$ stays constant at first and then increases rapidly as Q increases. Thus $(d_1(Q) + e_1(Q))/f(Q)$ will be very large for both small and large values of Q (Figure 4.7). Since $h(X)$ is a decreasing function of X (Figure 4.6), $X = h^{-1}\{[d_1(Q) + e_1(Q)]/f(Q)\}$ must be a decreasing function of h. Also, because h is large for both large and small values of Q, X must be small for both very small and very large values of N, being unimodal in between (Figure 4.8).

Finally, one can use the equilibrium condition for Equation (4.10a) to obtain

$$X = (I_n - r_n N)/(r_1 N - d_1 Q)$$

or, from Equation (4.11)

$$X = \{I_n - [d_1(Q) + e_1(Q)]Qr_n/r_1\}/[e_1(Q)Q] \tag{4.14}$$

When plotted for a variety of values of nutrient input I_n, (I_{n1} to I_{n7} in order of increasing nutrient input), Equation (4.14) has the family of curves shown in Figure 4.9.

The intersections of the curves from Equations (4.13) and (4.14) are the non-trivial equilibria (Q^*, X^*) of the system. When no intersection occurs, such as for curves designated by I_{n1} and I_{n7} in Figure 4.10, tree biomass

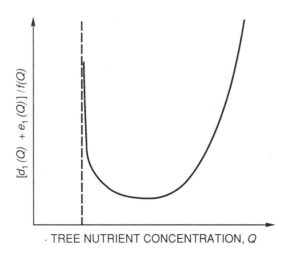

Figure 4.7   Plot of the mortality rate divided by efficiency of production of the autotroph as a function of internal nutrient concentration, Q, from Equation (4.11). (Based on Gatto and Rinaldi, 1987.)

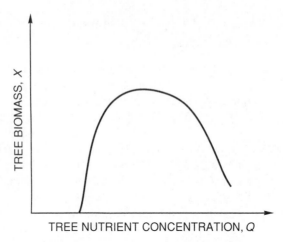

Figure 4.8   Tree biomass, X, as a function of internal nutrient concentration, Q. (Based on Gatto and Rinaldi, 1987).

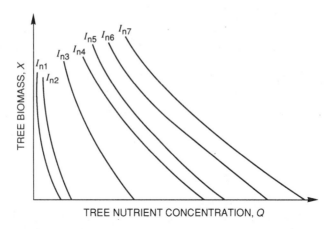

Figure 4.9   Family of curves of tree biomass, X, versus tree internal nutrient input, Q, for seven different values of nutrient input, I_n, from Equation 4.12. (Based on Batto and Rinaldi, 1987.)

does not exist in the steady state; there is either not enough nutrient to support the system or so much nutrient that there is a toxic effect on the tree, preventing the growth of biomass. Note that for values of I_n between I_{n4} and I_{n6} there are two intersections. It can be shown that only the equilibrium with higher tree biomass is stable; the other point is unstable. A critical situation exists when $I_n = I_{n6}$. The two steady-state equilibria have coalesced into one. Any further increase in I_n causes this point to disappear, representing tree extinction. This type of behaviour explains why

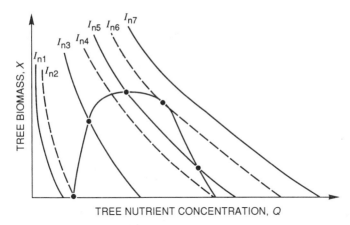

Figure 4.10   Intersections between the zero isoclines given by Equation (4.11) and those of Equation (4.12) for several values of I_n. The intersections represent possible steady-state equilibrium points. (Based on Gatto and Rinaldi, 1987.)

continuous changes in some parameters can sometimes cause discontinuous changes in variables.

The behaviour shown in Figure 4.10 is termed a 'fold catastrophe'. The reason for this name can be understood more clearly when the values of the steady-state tree biomass, X^*, are plotted as a function of nutrient input I_n (Figure 4.11). The solution curve representing biomass X^* as a function of nutrient input I_n folds back at the critical value I_{n6}, so there is no steady-state non-zero value of X^* for larger values of I_n. As I_n increases beyond I_{n6}, the system suddenly changes to a new state for which X^* goes to zero.

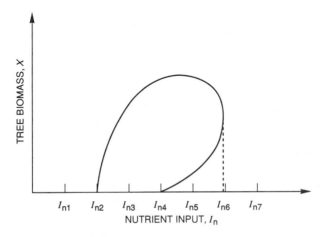

Figure 4.11   Tree biomass, X, as a function of nutrient input, I_n, showing a fold catastrophe at a rate of nutrient input of I_{n6}. (Based on Gatto and Rinaldi, 1987.)

The above example illustrates a large permanent change to a system resulting from a gradual environmental charge. Another type of situation that can occur is a permanent (or at least semipermanent) change to a system that can sometimes occur after a large perturbation. One example of such a change of state is that which can possibly follow a large perturbation of nutrient levels in a system. Stachurski and Zimka (1975), Gosz (1981), Vitousek (1982) and Pastor *et al.* (1984) have discussed changes in woody vegetation that might follow a disturbance greatly reducing the nitrogen availability of the soil. Woody plants might, under these circumstances, increase their nitrogen use efficiency, including the production of litter with low levels of nitrogen in relation to carbon. Mineralization of this litter could be very slow, as decomposers would immobilize the limiting nutrient. This nutrient immobilization would exacerbate the soil nutrient deficiency and cause plants to use the nitrogen more efficiently. Vitousek pointed out that the vegetation on the site might change phenotypically or genotypically as a long-term result of the perturbation, leading to a new steady state.

Evidence for the above type of phenomenon exists. Van Cleve *et al.* (1983) have studied two types of taiga autotroph communities, (1) black spruce type and (2) white spruce/hardwood type, that can, in principle, occupy some of the same geographic sites. They are characterized by different nutrient efficiencies, however. Black spruce have mechanisms that conserve nutrients, which is a useful feature for them, because the thick layers of organic matter that accumulate under black spruces effectively insulate the soil, reducing summer soil temperatures and slowing mineralization rates. White spruce/hardwood communities are not as efficient at nutrient cycling. It is possible that if a white spruce/hardwood system suffers a sharp reduction in nutrient availability, this deviation could be reinforced if black spruces were to invade and further reduce the availability of nutrients in the soil. A positive feedback effect could develop that would shift the system totally to black spruces.

4.8   FURTHER CONSIDERATIONS

The models described above are relatively simple treatments of the highly complex phenomena involved with internal nutrient dynamics in autotrophs. It is worth mentioning briefly some of the directions that more detailed modelling approaches have taken.

Collins (1980) described a phytoplankton model that had variables for external and internal levels of both nitrogen and phosphorus. The uptake functions for nitrate were generalized to include the inhibition by internal cell nutrient concentrations of further nutrient uptake (omission of this inhibition may have led to unrealistically large internal nutrient concentrations in the model simulations shown in Figure 4.4 and caused the catastrophe shown in Figure 4.10). Light and temperature effects on assimilation were also modelled. Collins calibrated the model with data from a subalpine lake and validated it with data from a geographically distant lake. She also tested a Monod-type growth model, which failed to describe the observed timing of the phytoplankton growth in the lakes.

Grenny *et al.* (1974) extended the internal nutrient approach further to model explicitly inorganic nitrogen, organic intermediate compounds, and protein within the phytoplankton cells. This internal complexity made it possible to assign to the phytoplankton a variety of different strategies of nitrogen uptake and allocation, allowing the cells to respond in different ways to variations in external conditions. Such diversity of possible strategies, which may change in relative advantage through time in a temporally fluctuating environment, may explain some of the observed diversity of phytoplankton species.

Further elaboration of physiological detail was incorporated into models such as those of Lehman *et al.* (1975) and Nyholm (1978), both of which are based on the internal nutrient store approach. By having detailed submodels for effects of light and thermal regimes and for other factors, such as settling or sedimentation of cells, these models were able to follow accurately seasonal population changes and successional changes of the phytoplankton community.

The internal dynamics of nutrients in multicellular terrestrial plants have also been the object of increasingly sophisticated modelling. As an example of one approach that has been especially useful in comparing different types of forests, Attiwill (1980) modelled a eucalypt forest, dividing biomass into leaves, branch wood, stembark, sapwood and heartwood, which change in relative proportions over the lifetime of a tree stand. The concentrations of nutrients also vary widely among the compartments (e.g. phosphorus concentration, averaged over a stand, is $0.716 \, \text{g kg}^{-1}$ dry weight in leaves but only $0.016 \, \text{g kg}^{-1}$ in heartwood). Attiwill noted an increased importance of biochemical and biogeochemical cycling as stand age increased.

Explicit models of nutrient and energy partitioning (e.g. between foliage, stemwood and roots) are useful for comparing strategies of plants under various conditions. McMurtrie (1986) used such a model to generate hypotheses concerning general principles of tree growth. For example, partitioning of energy to fine roots decreases sharply with increasing input of inorganic nutrient to the external available pool.

Similar to the case for phytoplankton models, a genre of highly detailed physiological models of tree and forest stand growth now exist (e.g. Dixon *et al.*, 1978; Luxmoore *et al.*, 1978; Ågren and Axelsson, 1980; Luxmoore, 1989). The first of these models, for example, couples a detailed soil chemistry model, two-solute uptake models, and a soil–plant–atmosphere water model. Models of this type may be especially useful in attempting to predict the effects of changing environmental conditions on forests.

4.9 SUMMARY AND CONCLUSIONS

This chapter has reviewed some results of generalizing the simple assumption that the autotrophic growth rate depends only on external nutrient levels. In more general models, such as the cell quota models of Droop (1968) and others, the nutrient concentration in an autotroph can vary through luxury consumption of the autotroph. The autotroph's growth rate depends on this internal concentration, rather than directly on ambient nutrient levels.

Although the cell quota models are conceptually more sophisticated than the simpler Monod-type growth model discussed in Chapter 3, they do not appear to predict fundamental differences in either the steady state or the stability behaviour of the model systems. The cell quota model for the system when the autotroph is the highest trophic level is locally stable for all reasonable parameter values (though this does not exclude the possibility that time lag effects between nutrient uptake and growth that are not built into these models may cause oscillations). Resilience properties of the cell quota model are also similar to the Monod-type growth rate dependence models of Chapter 3; that is the return time T_R does not vary appreciably with nutrient input I_n. However, the cell quota model also displays some fast time scale phenomena involving the rapid uptake of nutrient by cells.

A further complication in autotrophic dependence on nutrient levels is that nutrients may have toxic effects. When this is taken into account, together with rapid unrestricted uptake, in models, catastrophic behaviour can occur; that is, the steady-state equilibrium point for autotrophic biomass suddenly disappears when nutrient input to the system is increased beyond a critical level.

A variety of modelling approaches have extended the internal concentration approach to more detailed internal structure within the autotrophs and to more precisely described dependence on external environmental conditions. These increases in model sophistication allow one to study effects of changing external conditions and of the evolutionary strategies of the autotrophs.

5 Effects of nutrients on autotroph–herbivore interactions

5.1 INTRODUCTION

The addition of herbivory to the purely autotrophic systems studied in Chapters 3 and 4 is the next step in the progression from simpler to more complex nutrient cycling phenomena. There are two perspectives one can take at this stage. One perspective is to examine how the addition of herbivory alters the dynamics of the nutrient pool–autotroph–detritus systems already studied. The second perspective is to explore how the inclusion of nutrient cycling affects the dynamics of autotroph–herbivore systems. This second perspective is suggested by the fact that there is a vast body of research on autotroph–herbivore systems, generally without explicit consideration of the influence of a limiting nutrient.

The present chapter takes the second perspective, covering some of the ways in which the inclusion of nutrient limitation and recycling affect the autotroph–herbivore–detritus system. The first of these effects is on the nature of the steady state. Nutrient limitation can exert a 'bottom-up' control on a system that strongly influences the steady-state levels of the components. A second effect is on stability. This includes both local stability and resilience, or the speed with which a locally stable system approaches steady state following a perturbation. Some general considerations of herbivory and the way it is described in mathematical models are first discussed.

5.2 HERBIVORY

Two types of herbivore–autotroph systems have been the foci of particular attention. These are terrestrial plant-grazer systems, where the grazer is normally a mammal or insect (e.g. Mattson and Addy, 1975; Noy-Meir, 1975; Ludwig *et al.*, 1978; Myers, 1979; Crawley, 1983; Strong *et al.*, 1984) and aquatic or marine phytoplankton–zooplankton systems (e.g. Steele, 1974a, 1978; McNaught and Scavia, 1976; Kremer and Nixon, 1978; Powell and Richerson, 1985). A wide variety of modelling approaches has been used, reflecting the qualitative diversity of grazing systems.

Caughley (1976) divided grazing systems into two general types: 'non-interactive' and 'interactive'. A non-interactive type is one in which the grazer does not influence the rate at which an autotroph is renewed. The general equation for the autotroph's rate of change in this case is

$$dX/dt = G(X) - F(X, Y) \qquad (5.1)$$

where X and Y are autotroph and herbivore biomasses, respectively, and where

$G(X)$ = net production rate of the autotroph; a function of autotroph biomass X only, not of herbivore biomass Y,

$F(X, Y)$ = removal rate of autotroph biomass by the herbivore.

This model may be valid for many cases in which the grazer has no influence on the autotroph other than removal of biomass. The dynamics of such systems have been studied extensively by means of predator–prey or, more generally, consumer–resource models (e.g. Rosenzweig and MacArthur, 1963; Gallopin, 1971a,b; Murdoch and Oaten, 1975; DeAngelis et al., 1975).

The situation is quite different when nutrient cycling is taken into account, however. If the herbivore influences the rate of recycling of nutrients to the available pool, N, then it indirectly affects the growth rate of the autotroph. According to Caughley's (1976) categorization, then, an autotroph–herbivore system where nutrient recycling by the herbivore plays a major role should in general be called 'interactive'. The effects of this interaction will be studied more closely in Chapter 6, where the effects of herbivore recycling of nutrients is the main topic. For the topics of this chapter, recycling is not as important as the role nutrients play in limiting autotroph growth and affecting autotroph–herbivore dynamics.

The autotroph–herbivore interaction is represented in the removal rate, $F(X,Y)$, which normally depends on the densities of both the autotroph and herbivore populations. A number of different phenomenological functions have been proposed to represent the herbivore feeding rate, including

(1) $F(X,Y) = fXY$	(Lotka–Volterra)
(2) $F(X,Y) = fXY/(b + X)$	(Holling Type II)
(3) $F(X,Y) = fY(1 - e^{-kX})$	(Ivlev)
(4) $F(X,Y) = fX^2Y/(b + X^2)$	(Holling Type III)
(5) $F(X,Y) = fX^2Y/[(b_1 + X)(b_2 + X)]$	(Jost et al., 1973)
(6) $F(X,Y) = f(X - X_0)Y/(b + X)$	(Holling Type II with threshold)
(7) $F(X,Y) = fX^{1/2}Y/(b + X^{1/2})$	(Gause)

where f, b, k, b_1, b_2 and X_0 are constant parameters. Note that the Holling Type II function has the same general form as the Monod-type growth function for nutrient–autotroph interactions used earlier. Arguments for justifying the use of this type of function for consumer–resource interactions between organisms were made by Holling (1959), Cushing (1959, 1968) and Crawley (1973). In particular, it approximates a great deal of feeding information very well. All of the functions except the Lotka–Volterra function share the feature that, as the autotroph biomass reaches very high levels, the dependence of $F(X,Y)$ on X weakens, and, ultimately diminishes towards zero as $X \to \infty$, so that $F(X,Y)$ virtually becomes a function only of Y, the herbivore biomass. This is termed a 'saturation' effect of X, representing the fact that there is a physiological limit to the rate at which autotroph biomass can be consumed and assimilated per unit herbivore biomass. Although the Lotka–Volterra model is less realistic because it does not take the saturation effect into account, it is a helpful approximation in some cases where its simplicity makes analysis easier to perform. It will be used later in some cases, though conclusions based on it (or any of these simple functions) must be interpreted with caution.

The Holling Type III function, which is quite similar to the Jost function, is similar to the Holling Type II function in embodying the effect of saturation, but it tends to lead to more stable consumer–resource interactions than does the Holling Type II function (e.g. Nunney, 1980). Note that the Type III function has an X^2 in the numerator, which means that the rate of consumption per unit autotroph biomass, F/X, decreases as X decreases. This function represents cases in which a given unit of resource biomass becomes more difficult for herbivores to obtain as total prey biomass decreases. (In contrast, the Gause function predicts that the consumer becomes more effective at obtaining resources at low resource densities.) The type of behaviour described by the Holling Type III function can easily happen if some fraction of the resource biomass is relatively well protected from consumption, by the existence of resource refuges, for example. In the case of autotrophs, these 'refuges' may include physical locations that are difficult to find, but they can also include other factors that make some fraction of the population more difficult to consume than the rest for some reason. Tall trees, for example, are less vulnerable to herbivores than are seedlings or saplings. The resources which do not have safe refuges are the first to be captured by consumers, so that the resources that are left are the ones that are harder to find or capture (Rosenzweig and MacArthur, 1963). Alternatively, the Holling Type III function can indicate that prey in low abundances are harder to find and capture because the consumer is less apt to have a 'search image' for the prey. The Holling Type III function has proven relatively successful in describing feeding by vertebrates, but the Type II function is more successful for describing feeding by insects.

Reviews of consumer–resource dynamics can be found in Taylor (1984) and elsewhere. It will be useful, however, to discuss some important features of these dynamics here.

If nutrient cycling and the detrital compartment are ignored for the moment, the interaction between a consumer and its resource can be written as

$$dX/dt = G(X) - F(X,Y) \tag{5.2a}$$

$$dY/dt = \eta F(X,Y) - H(Y) \tag{5.2b}$$

where η ($\eta < 1$) represents the efficiency of biomass conversion, and $H(Y)$ is the loss rate of herbivore biomass. To take a specific case, let

$$G(X) = (r_1 - g_1 X)X \tag{5.3a}$$

$$H(Y) = d_2 Y \tag{5.3b}$$

where r_1, g_1 and d_2 are constants. Consider in turn both the Lotka–Volterra and Holling Type II functions for $F(X,Y)$. Typical state plane diagrams for these two models are shown in Figure 5.1. In Figures 5.1(a,b) the zero isoclines ($dX/dt = 0$ and $dY = 0$) are drawn for two cases: the Lotka–Volterra interaction term with $g_1 > 0$ and $g_1 = 0$ respectively. In Figures 5.1(c,d), the zero isoclines are drawn for two particular cases of the Holling

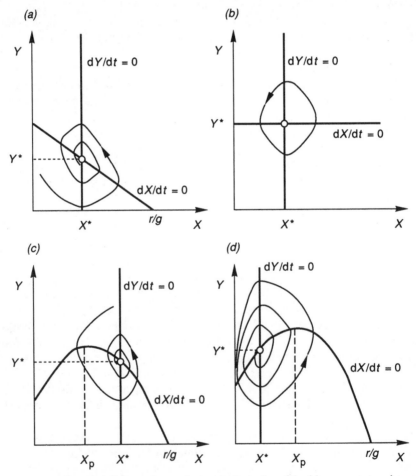

Figure 5.1   Zero isoclines and population trajectories of herbivore–autotroph systems for different assumptions concerning the interactions between the two trophic levels. See text for discussion.

Type II interaction. In all four cases, typical trajectories are drawn. To plot the zero isoclines, the right-hand sides of Figure 5.2(a,b) must be set to zero and solved for Y in terms of X, or vice versa. For the case of the Lotka–Volterra functions, Equation 5.2(a) yields $Y = (r_1 - g_1 X)/f$ and Equation 5.2(b) yields $X = d_2/(\eta f)$.

Note that the Lotka–Volterra model in Figure 5.1(a), where $g_1 > 0$, is stable. That is, for every set of starting conditions (X_i, Y_i) of the variables, the trajectory ultimately approaches the equilibrium point (X^*, Y^*), at the intersection of the two zero isoclines. This specification of $g_1 > 0$ represents the case when the carrying capacity of the environment for the autotroph in the absence of herbivory, which is r_1/g_1, is finite. When $g_1 = 0$, however, which represents an infinite carrying capacity, the resulting system is referred to as being only 'neutrally' stable. If the initial values were set precisely to (X^*, Y^*), the system would remain at that point. Any slight deviation from this point, however, would result in the variables oscillating indefinitely along some closed trajectory, such as that in Figure 5.1(b). A larger perturbation away from (X^*, Y^*) would result in an oscillation with a larger amplitude (May, 1973).

When the Lotka–Volterra feeding function is replaced by a Holling Type II function, the dynamics differ from that of the Lotka–Volterra system in an important way. The $dX/dt = 0$ isocline (which the reader can verify is described by $Y = (r_1 - g_1 X)(b_1 + X)/f$) now can have a 'hump' for a range of parameter values, and the system can exhibit either stable or unstable behaviour, depending on what side of the hump the equilibrium point lies (Figures 5.1c,d). It is useful to illustrate briefly why this is so, using a derivation similar to that which Gilpin (1972) used.

When the Holling Type II function $F(X,Y) = fXY/(b_1 + X)$ is substituted into Equations (5.2a,b), where $G(X)$ and $H(Y)$ are given by Equations (5.3a,b), the peak of the prey zero isocline can be found by the procedure of solving for a maximum value of a function. In this case the function is the expression for the zero isocline; $Y = (r_1 - g_1 X)(b_1 + X)/f$. This is done by solving $dY/dX = 0$, at which point Y is an extremum (it is also a maximum if it is true that $d^2Y/dX^2 < 0$). The peak occurs at the point $X = X_p$, where

$$X_p = (r_1 - g_1 b_1)/2g_1 \tag{5.4}$$

The equilibrium value of X, X^*, occurs where the two zero isoclines, $dX/dt = 0$ and $dY/dt = 0$, intersect. In this case, since the zero isocline dY/dt is a vertical line, this equilibrium value of the autotroph is found by solving $dY/dt = [\eta fXY/(b_1 + X)] - d_2 Y = 0$ for X^*;

$$X^* = d_2 b_1/(\eta f - d_2) \tag{5.5a}$$

while the equilibrium value of the herbivore, Y^*, is found by setting $X = X^*$ in the expression for the $dX/dt = 0$ isocline, found earlier. Then

$$Y^* = (r_1 - g_1 X^*)(b_1 + X^*)/f \tag{5.5b}$$

It can be shown now that if $X^* > X_p$, the autotroph–herbivore system is stable, whereas, if $X^* < X_p$, the system is unstable. This can be proven by analysing the equations obtained by linearizing Equations (5.2a,b) about (X^*, Y^*) [through substitution of $X = X^* + X'$ and $Y = Y^* + Y'$, as in Equations (3.22) of Chapter 3]:

$$dx'/dt = [-g_1 X^* + fX^*Y^*/(b_1 + X^*)^2]x' - fX^*y'/(b_1 + X^*) \qquad (5.6a)$$

$$dy'/dt = \eta f X^* Y^* x' b_1/(b_1 + X^*)^2 \qquad (5.6b)$$

The two eigenvalues of this system are determined from the equation

$$\det \begin{vmatrix} -g_1 X^* + fX^*Y^*/(b_1 + X^*)^2 - \lambda & -fX^*/(b_1 + X^*) \\ \eta f X^* Y^* b_1/(b_1 + X^*)^2 & -\lambda \end{vmatrix} = 0$$

or, after solving for

$$\lambda\pm = -\tfrac{1}{2}[g_1 X^* - fX^*Y^*/(b_1 + X^*)^2] \pm \tfrac{1}{2}\{[g_1 X^* - fX^*Y^*/(b_1 + X^*)^2]^2 - 4\eta f^2 X^{*2} Y^* b_1/(b_1 + X^*)^3\}^{1/2} \qquad (5.7)$$

There can be a positive real part for one of the roots, $\lambda\pm$, if and only if

$$g_1 X^* - fX^*Y^*/(b_1 + X^*)^2 < 0 \qquad (5.8)$$

When Y^* from Equation (5.5b) is substituted into (5.8), the inequality becomes

$$g_1(b_1 + X^*) < r_1 - g_1 X^*$$

or

$$X^* < (r_1 - g_1 b_1)/2g_1 = X_p \qquad (5.9)$$

Thus, the autotroph–herbivore system is unstable when inequality (5.9) holds, implying, from Equation (5.4), that the system is unstable when $X^* < X_p$ and stable otherwise. May (1972) proved that not only for the consumer–resource system represented by Equations (5.2a,b) and (5.3a,b), but for many others that are commonly used, the system can be unstable under certain conditions and gives rise to limit cycle oscillations, as shown in Figure 5.1d. These are oscillations in which the consumer and resource both oscillate with the same periodicity. They differ from the neutral oscillations of Figure 5.1(b). Any perturbation of the neutral oscillations would cause the consumer–resource trajectory to change to a new orbital trajectory, whereas no matter how the consumer and resource are perturbed in Figure 5.1(d), the trajectory eventually returns to the same limit cycle trajectory.

An example of a herbivore–autotroph interaction that appears to be characterized by a limit cycle is that of the snowshoe hare and woody plants in Alaska (Fox and Bryant, 1984). The woody plants are destructively overbrowsed during peaks in hare populations and may take a few to several years to recover their palatability. The hare population 'crashes' in the interim.

Because it is the positive slope in the prey isocline (i.e. the isocline $dX/dt = 0$) to the left of its peak that is associated with the instability of the consumer–resource model of Figures 5.1(c,d), one way of eliminating this instability is to find ways to change the system so that this slope is never positive (see Rosenzweig and MacArthur, 1963). The Holling Type III function, $F(X,Y) = fX^2Y/(b_1 + X^2)$, which can be justified ecologically as a reasonable interaction in many cases, tends to eliminate the peak and stabilize the system. As discussed earlier, when X decreases to very small values, $F(X,Y)/X$, the per unit biomass loss rate of resource, decreases. This can be interpreted as meaning that the lower the resource level is, the more secure the remaining resource is from consumption. This changes the shape of the prey zero isocline (Figure 5.2) and in this case makes the autotroph–herbivore system completely stable. (In general not every case of a Holling Type III interaction is stable, but compared with a Holling Type II function, this function at least decreases the range of X^* over which the system is unstable.) The feeding function, $F(X,Y) = f(X - X_0)Y/(b_1 + X)$, which is the Holling Type II function with a threshold on the prey level needed for herbivore growth, also tends to be more stable than the Holling Type II function (which is the special case when $X_0 = 0$).

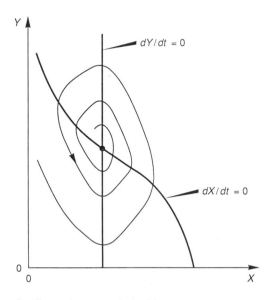

Figure 5.2   Zero isoclines of autotroph–herbivore systems when a Holling Type III feeding function is used. Compare with the zero isoclines associated with the Holling Type II feeding function in Figures 5.1(c,d).

5.3   EFFECTS OF NUTRIENT INPUT ON STEADY-STATE PATTERNS

Now these well-known autotroph–herbivore dynamics can be placed in the context of nutrient cycling. The simplest assumption is that the dynamics of the autotroph–herbivore interaction has no feedback effect on nutrient level, N, but that the level of nutrient controls the rate of autotroph growth; that is, the growth rate r is written as $r_1 = r(N)$, where $r(N)$ is a non-decreasing function of N. An example is the familiar Monod-type growth function, $r(N) = r_1 N/(k_1 + N)$ (see Chapter 3).

Consider the system with a Holling Type II feeding function for herbivory, so that the equations for autotroph and herbivore biomass are

$$dX/dt = r(N)X - g_1 X^2 - fXY/(b_1 + X) \tag{5.10a}$$

$$dY/dt = [\eta fXY/(b_1 + X)] - d_2 Y \tag{5.10b}$$

for a given nutrient level, N. Successively increasing values of N cause the autotroph zero isocline, $Y = [r(N) - g_1 X](b_1 + X)/f$, to change, as shown in Figure 5.3. For the lowest value, $N = N_1$, the equilibrium is $X^* = r(N_1)/g$, $Y^* = 0$, meaning that the herbivore cannot be supported by that low a nutrient level. For a higher nutrient level, $N = N_2$, $Y^* = Y_2$, and for the highest level $N = N_4$, $Y^* = Y_4$. It can be shown generally that Y^* increases monotonically as N^* increases. Clearly, the nutrient level has an important effect on the herbivore biomass, Y^*. At the same time, the

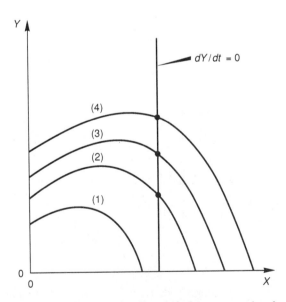

Figure 5.3   Changes in the zero isoclines of the autotroph of an autotroph–herbivore system as the nutrient level N increases in Equations 5.10(a,b).

autotroph level remains fixed at $X^* = b_1 d_2/(\eta f - d_2)$ in all cases. This is an example of a 'top-down' effect from the highest trophic level, which determines the level of the next lower trophic level. This top-down control is an implicit assumption of the model [Equation (5.10)] and it appears to fit some natural food chains, as discussed later.

It may be more interesting to explore this pattern of change in greater detail for a specific model. Since it is easier to analyse, let us start with a Lotka–Volterra interaction between the autotroph and the herbivore. When embedded into an environment that includes a recycling nutrient and a detrital component (Figure 5.4), the autotroph–herbivore system can be described by the equations for nutrients, autotroph biomass, herbivore biomass and detrital biomass per unit area;

$$dN/dt = I_n - r_n N - [\gamma r_1 NX/(k_1 + N)] + \gamma d_D D \tag{5.11a}$$

$$dX/dt = [r_1 NX/(k_1 + N)] - fXY - (d_1 + e_1)X \tag{5.11b}$$

$$dY/dt = \eta fXY - (d_2 + e_2)Y \tag{5.11c}$$

$$dD/dt = (1 - \eta)fXY + d_1 X + d_2 Y - d_D D \tag{5.11d}$$

where $r(N)$ is now given the particular form of a Monod-type growth function, $r_1 N/(k_1 + N)$. The term $-g_1 X^2$, which occurred in Equation (5.3a) and represents the carrying capacity limitations on the autotroph, has been dropped from the equation for X, it being assumed now that the nutrient availability alone will set the ultimate carrying capacity for the autotroph.

The constant γ is the ratio of nutrient to biomass, which is assumed, for simplicity, to be the same for the autotroph and herbivore. While this

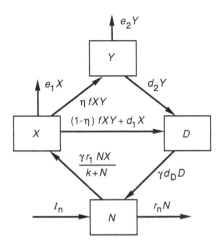

Figure 5.4 Autotroph–herbivore system ($X–Y$) embedded in an ecosystem with a nutrient pool N and detritus D. Arrows represent nutrient or biomass flows.

assumption is unrealistic, it does not affect the qualitative conclusions that we desire at this point. The coefficients d_1 and d_2 refer to mortality rates of autotrophs and herbivores associated with retention of the biomass in the system, while e_1 and e_2 are coefficients of the rate at which the biomass is lost from the system, by drift, for example. The coefficient η is the fraction of biomass removed from the autotroph by consumption that is assimilated by the herbivore. The rest goes straight to the detrital compartment.

The steady-state values of the variables are, when all coexist, given by

$$N^* = -(1/2)b \pm (1/2)(b^2 - 4c)^{1/2} \tag{5.12a}$$

$$b = k_1 - \frac{I_n}{r_n} - \left(\frac{d_2}{f} - X^*\right)\frac{\gamma r_1}{r_n} - \frac{\gamma d_1 X^*}{r_n} + \frac{\gamma d_2}{f r_n}(d_1 + e_1)$$

$$- \frac{\gamma(1-\eta)}{r_n}(r_1 - d_1 - e_1)$$

$$c = \frac{kI_n}{r_n} - \frac{\gamma d_1 X^* k}{r_n} + \left[\frac{\gamma(1-\eta)fX + \gamma d_2}{f r_n}\right](d_1 + e_1)k_1$$

$$X^* = (d_2 + e_2)/(\eta f) \tag{5.12b}$$

$$Y^* = (I_n - r_n N^* - \gamma e_1 X^*)/(\eta \gamma f X^* - \gamma d_2) \tag{5.12c}$$

$$D^* = (d_1 X^* + d_2 Y^* + (1-\eta)fX^*Y^*)/d_D \tag{5.12d}$$

Note from Equation (5.12c) that for Y^* to be greater than zero, the inequality $I_n > r_n N^* + \gamma e_1 X^*$ must be true. Thus, there is a threshold value of nutrient input for herbivores to be maintained in this model system. A striking aspect of the solution is that when herbivory is present in the system, X^* is independent of the nutrient input I_n; that is, the autotroph is entirely controlled from the top down. It is interesting to plot X^*, Y^* and N^* over the whole range of values, starting from $I_n = 0$ (Figure 5.5). The other parameter values are the same as those used in Figure 3.7 (from O'Brien, 1974), with the additional parameters, representing herbivory, as follows: $f = 10^{-8}$, $d_1 = 0.0$, $d_2 = 0.0$, $e_1 = 0.20 \, \mathrm{d}^{-1}$, $e_2 = 1.01 \, \mathrm{d}^{-1}$, $d_D = 0.0$, $\eta = 0.5$. Because X and Y are interpreted as whole organisms, not biomass in O'Brien's model, the product of constants, ηf, is a rate of assimilation of phytoplankton cells into zooplankters in units (phytoplankton)$^{-1} \mathrm{d}^{-1}$. If the average size of zooplankters is such that assimilation of many phytoplankters is necessary to produce a single zooplankter of average size per unit time, it is necessary to scale Y in Equations 5.11(b,c,d) by ρ, where ρ is the ratio of individual zooplankter biomass to individual phytoplankter biomass, the preserve balance of nutrient flux. Here we will assume $\rho = 100$. This represents a system with no nutrient recycling back to the nutrient pool, N. (The effects of recycling on this system are considered in the next chapter.) For consistency with O'Brien's variables, X is now assumed to represent the number density of phytoplankton cells rather than biomass,

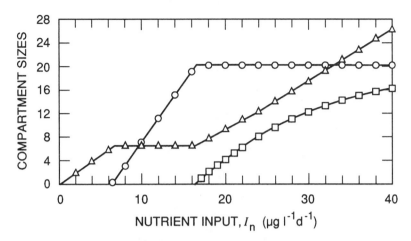

Figure 5.5 Changes in the nutrient amount N in water ($\mu g \, l^{-1}$) (triangles), number of phytoplankton cells ($\times 10^7 \, l^{-1}$) (circles) and herbivore (zooplankton) cells ($\times 10^4 \, l^{-1}$) (squares) as functions of nutrient input, I_n; described by Equations (5.12a,b,c,d).

and Y can be taken as the number density of herbivores (zooplankters, for example, though no attempt is made to represent any particular real species). The sequence of changes for increasing I_n can be explained as follows. For very small values of I_n, both $X^* = 0$ and $Y^* = 0$, so N^* increases directly in proportion to I_n; $N^* = I_n/r_n$. When nutrient input exceeds the threshold I_{n1},

$$I_{n1} = r_n k_1 (d_1 + e_1)/(r_1 - d_1 - e_1), \tag{5.13}$$

the autotroph can be supported, as described in Chapter 3. Autotroph numbers increase linearly with I_n until a threshold where the herbivore can invade is passed at the point where $\eta f X^* > d_2 + e_2$, which occurs when nutrient input level reaches the level I_{n2};

$$I_{n2} = (\gamma e_1/\eta f)(d_2 + e_2) + r_n k_1 (d_1 + e_1)/(r_1 - d_1 - e_1) \tag{5.14}$$

After this threshold of I_n is exceeded, N^* increases in proportion to I_n, while X^* is held constant by the herbivore. The herbivore number density, Y^*, increases at first, but then gradually asymptotes to a constant value. The reason that Y^* does not increase indefinitely is the saturation effect that limits the total flux of energy and nutrient moving up the food chain.

This saturation effect relates to the general question of top-down versus bottom-up control, which will be discussed more thoroughly in Chapter 8. In the present case the herbivore, when present, controls the autotroph number density (or, equivalently, standing stock biomass). Studies of biotic communities have provided numerous examples of changes in the abundance and composition of lower trophic levels resulting from changes in the numbers of grazers and predators (e.g. Paine, 1966). Evidence of food chain control of plankton community structure came initially from observations of changes in algal density (evidenced by water turbidity) and size distributions of zooplankton after removal of fish from whole lakes or ponds. *Daphnia*, in particular, have been noted as capable of controlling natural algae in ponds and lakes (e.g. Hrbacek *et al.*, 1961; Shapiro, 1980; Lynch and Shapiro, 1980; Edmondson and Litt, 1982). The present model seems to indicate that the herbivores will control the autotrophs even under very eutrophic conditions, although this depends on the autotrophs being edible forms. Experimental manipulations of Levitan *et al.* (1985) of a small lake by removals of trout that feed on *Daphnia* indicate that the *Daphnia* under those circumstances are capable of responding quickly to nutrient-stimulated algae and limiting algal biomass. The effects of higher trophic levels on this autotroph–herbivore system are considered in Chapter 8.

Elliott *et al.* (1983) tested the hypothesis that phytoplankton were herbivore-controlled in even-length food chains (the autotroph–herbivore chain is even in length). They found low levels of algal standing stocks and high levels of soluble nutrients under these circumstances, similar to the present model results.

A number of studies indicate increases in herbivore biomass as nutrient input is increased. Hershey *et al.* (1988) enriched a tundra stream with phosphorus and noted increases in the growth and density of a tube-building larval chironomid downstream. Harper (1986) noted increases in biomass densities of zooplankton from less to more nutrient-enriched lochs in Scotland. Intertidal and nearshore zones enriched with nutrients from coastal upwellings supported higher densities of limpets, which graze on algae (Bosman *et al.*, 1987). Wroblewski *et al.* (1988) simulated phytoplankton–zooplankton dynamics in the North Atlantic as a function of total nutrients and obtained results similar in form to Figure 5.5, with zooplankton becoming established above some threshold in N and increasing towards an asymptote. Comparisons with zooplankton abundance in the North Atlantic showed that the model predictions were too high, though the authors suggested that the discrepancy might be due to an overly large assumption of total nutrients in the model.

5.4 EFFECTS OF NUTRIENTS ON AUTOTROPH–HERBIVORE STABILITY

We have seen above that the nutrient input level, I_n, of the autotroph–herbivore system affects the structure of steady-state solutions. One may also ask how it affects the stability of the system. By stability here we mean local or Lyapunov stability, defined in Chapter 2 as the tendency of all system components (or variables in mathematical models of the systems) to return to their steady-state equilibrium values following small perturbations that move them away from these values, as shown in Figure 5.1(*c*). Local stability has long been a topic of interest for theoretical ecologists because instability can lead to large population oscillations and extinctions. If the mathematical model of an ecological community is locally stable, it indicates the tendency towards constancy of populations. Thus local stability is related to the question: Why do certain food web structures exist as relatively stable steady states, whereas others exhibit undamped oscillations or rapid species extinctions? A central finding of the 1970s was that large food web models tend to be unstable and that the probability of instability increases with increasing numbers of components and interconnections between these components (May, 1972, 1973).

Another key result of the 1970s was Rosenzweig's (1971) prediction of the 'paradox of enrichment', which indicated that, in a system with a nutrient-limited autotroph population and a herbivore population, decreasing the level of nutrient input decreased the possibility of oscillations or extinction of one or both species. Rosenzweig's result applies to models with interactions such as the Holling Type II function, which is illustrated in Figures 5.1(c,d). Such functions for consumption are called 'depensatory', because the consumption rate per capita prey, $fY/(b + X)$, decreases as prey density increases. This depensatory effect is destabilizing because, if the autotroph population is rapidly growing, it can temporarily escape or 'overshoot' the control of the herbivore and continue to grow until other limitations arrest autotroph growth. As the autotroph population growth slows, the herbivore population catches up, producing a population crash of the autotrophs, followed by a crash of the herbivore population. These so-called limit cycle oscillations can continue indefinitely unless one or more of the populations in the system becomes extinct.

Rapid autotroph growth enhances the depensatory instability of the autotroph–herbivore system. Theoretical results of Rosenzweig (1971) demonstrated that nutrient limitation, by restricting the growth rate of the autotroph, helps stabilize autotroph–herbivore systems and large food webs. Following up on this idea, theoreticians investigated further the stability of food chains and webs, with both open and closed nutrient cycles for a variety of trophic interaction functions.

To review from Chapter 1, an open system with respect to nutrient flow is one in which there are inputs to and outputs from the system, as in Figure 5.4. In a closed system, there are no inputs or outputs, and the total amount of nutrient in the system is conserved. Some of the model studies have been done assuming open systems, while others have assumed systems that are self-contained or closed to nutrient input or output.

Parker (1978) started with an autotroph–herbivore system with a Lotka–Volterra interaction and no nutrient limitation. That system was unstable. When he extended the model by adding a nutrient pool and nutrient-limited growth of the autotroph (assuming a closed nutrient cycle), the periodic cycles of the autotrophs and herbivores damped out in all cases.

Model studies have included a variety of other trophic interaction functions besides Lotka–Volterra terms. Systems with depensatory interactions, such as the Holling Type II, may be unstable even in closed nutrient cycle models. However, decreases in the total nutrient in the system, N_T, eventually move the systems into domains where only stable steady-state solutions exist (Steele and Henderson, 1981).

Sjoberg (1977) considered a closed nutrient cycle model (available nutrient pool, autotrophs, herbivores and detritus) and either linear donor-dependent or Holling II interactions between components. All combinations of functions were stable except when both primary production and grazing functions were Holling Type II. In this case, if the model autotroph was assumed to return some of its stored nutrient directly to the detritus, stability of the system was increased above that of a similar system without such a direct pathway (studied by Steele, 1974a).

The systems described above were all closed with respect to nutrient fluxes. Theorists have also modelled nutrient limitations in open systems. Bader *et al.* (1976) modelled an autotroph–herbivore system (blue–green algae and a protozoan) in a chemostat, along with the nutrient substrate utilized by the autotroph prey. No feedback of nutrient substrate from the autotroph or herbivore to the nutrient pool was assumed. Three different herbivore growth functions (exhibited earlier in this chapter) were used: the Holling Type II, the multiple saturation model (similar to Holling Type III), and the Gause model. The different models had significant differences in their stability properties. In particular, the Holling Type III function tended to confer stability. In all cases, however, a decrease in the nutrient input pushed an initially unstable system towards a stable steady state. (Decreasing the nutrient input too far, however, could cause herbivore extinction.)

An example will show how the limitation of nutrient input can stabilize an autotroph–herbivore system with a Holling Type II function for herbivore grazing and a Monod-type growth function for autotroph growth. In the absence of the consideration of nutrients (and disregarding detritus here), the equations governing the biomasses X and Y of the autotroph–herbivore system are

$$\mathrm{d}X/\mathrm{d}t = r_1 X - [fXY/(b_1 + X)] - d_1 X - g_1 X^2 \tag{5.15a}$$

$$\mathrm{d}Y/\mathrm{d}t = [\eta fXY/(b_1 + X)] - d_2 Y \tag{5.15b}$$

where the equilibrium point is assumed to lie to the left of the peak of the autotroph hump, as shown in Figure 5.1(*d*). When a limiting nutrient pool and a detrital compartment are added to the model, the nutrient-limited autotroph production can be modelled as $r_1 XN/(k_1 + N)$. The equations for the system are

$$\mathrm{d}N/\mathrm{d}t = I_n - r_n N - [\gamma r_1 XN/(k_1 + N)] + \gamma d_D D \tag{5.16a}$$

$$\mathrm{d}X/\mathrm{d}t = r_1 XN/(k_1 + N) - [fXY/(b_1 + X)] - (d_1 + e_1)X - g_1 X^2 \tag{5.16b}$$

$$dY/dt = \eta f XY/(b_1 + X) - (d_2 + e_2)Y \tag{5.16c}$$

$$dD/dt = [(1 - \eta)f XY/(b_1 + X)] + d_1 X + d_2 Y + g_1 X^2 - d_D D \tag{5.16d}$$

where e_1 and e_2 represent biomass losses from the system (loss from detritus is ignored here). This is an open system with regard to nutrients; there is an input I_n and loss rates of nutrients associated with the biomass loss rates from the system as well the loss rate of nutrient directly from the system, $r_n N$. It is assumed that there is a density-dependent loss of autotroph biomass, $g_1 X^2$, to the detritus.

In Figure 5.6(a,b) plots of X and Y versus time for two different values of I_n are shown, indicating that small enough values of I_n can stabilize the system. This phenomenon is an example of Rosenzweig's (1971) 'paradox of enrichment', which, as stated earlier, noted that an increase in available nutrients that allows autotrophs to grow faster can destabilize an autotroph–herbivore system.

To what extent have theoretical findings that relate nutrient limitation and stability been borne out by empirical observations? The experimental evidence from field and laboratory studies suggests that increased concentrations of nutrients can have a destabilizing effect on phytoplankton–zooplankton systems, often leading to limit-cycle behaviour. For example, *Daphnia magna* and phytoplankton maintained in indoor 3400-litre tanks exhibited stable coexistence under low nutrient loading conditions, but the system became unstable when subjected to phosphorus additions greater than a defined limit (Borgmann *et al.*, 1988). In smaller laboratory systems, Luckinbill (1974, 1979) found that the stability of food webs of ciliate predators, *Didinium nasutum*, and their prey depended critically on nutrient levels; enrichment led to violent predator–prey cycles and to extinction. Nutrient (nitrate) limitation appears to have damped oscillations of rotifers and algae in a continuous culture (Boraas, 1980).

In field studies, additions of nutrients to enclosures containing lake water led to large increases in phytoplankton concentrations that were usually closely followed by population crashes (e.g. Hessen and Nilssen, 1986). In the unfertilized treatment, phytoplankton populations were more stable. Observations of tightly coupled oscillations in the numbers of bacteria and flagellates (Anderson and Fenchel, 1985; Anderson and Soerensen, 1986; Fenchel, 1982) indicated the potential for limit cycles induced by nutrient enrichment in heterotrophic food webs as well. An experimental addition of nutrients increased the amplitude of oscillations of bacteria and heterotrophic flagellates grazing on the bacteria in seawater enclosures studied by Bjornsen *et al.* (1988).

The effects of nutrient enrichment on stability are not as well documented for terrestrial systems as they are for aquatic systems, probably because of the longer time scales necessary for studies in terrestrial systems. As a result of nutrient enrichment, plant communities in terrestrial systems have shown

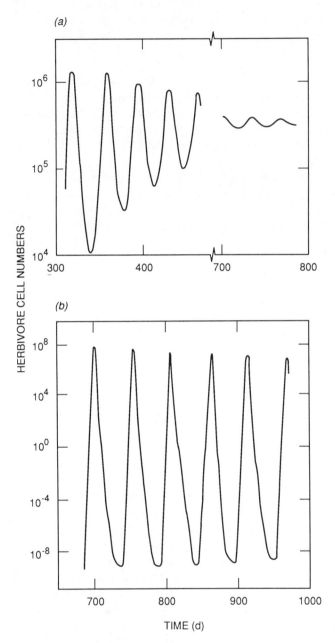

Figure 5.6 Herbivore cell numbers predicted by Equations (5.16a,b,c,d). (*a*) Nutrient input $I_n = 10\,\mu g\,l^{-1}\,d^{-1}$ and (*b*) $I_n = 200\,\mu g\,l^{-1}\,d^{-1}$. Case (*a*) is clearly stable. Case (*b*) may be converging over the long term, but is effectively unstable over short time scales, indicating the destabilizing effects of enrichment. Other parameter values are $r_n = 1.0$, $\gamma = 2.36 \times 10^{-7}$, $r_1 = 0.5$, $k_1 = 10$, $k_2 = 10$, $d_1 = 0.15$, $d_2 = 1.0$, $e_1 = 0.05$, $e_2 = 0.01$, $\eta = 0.05$, $f_1 = 10^{-7}$, $g_1 = 0.0$.

short-term increases in biomass and production (e.g. Bakelaar and Odum, 1978; Mellinger and McNaughton, 1975), but no long-term oscillations have been observed.

Increased nutrient availability can sometimes decrease the stability of terrestrial food webs by releasing herbivores from nutrient limitation, potentially leading to overexploitation of the food supply. It is well known that the level of nutrients in a plant can be highly variable (Chapter 4 and Chapin *et al.*, 1980), reflecting levels of nutrients available in the environment. If the nutrient is limiting to the herbivore, changes in plant nutrient content can affect survival and reproduction of the herbivore (e.g. Sedlacek *et al.*, 1988), potentially affecting the stability of the interaction. In reviewing the effects of plant nitrogen on herbivores, Mattson (1980) noted that survival rates of sucking insects typically increase with nitrogen content. Studies have shown that some chewing insects suffer direct decreases in survival on plants with high levels of nitrogen, but that the indirect effects of lower N are generally negative for herbivore growth and survival. In fact, low plant nutrient levels may be a strategy against herbivory (Chapin *et al.*, 1980).

White (1974) suggested that insect outbreaks could occur when plants have more nutritional quality, such as a higher nitrogen content. The same has been proposed for herbivorous mammals (e.g. Schultz, 1969). The higher grazing levels accompanying an outbreak can lead to severe reduction in foliage and in its nutrient quality. According to some hypotheses (Kalela, 1962; Pitelka, 1964; Schultz, 1964; Tast and Kalela, 1971; Laine and Henttonen, 1982; Haukioja *et al.*, 1983), plants may take several years to recover from heavy grazing, during which time herbivore populations may crash.

Myers (1980) hypothesized that high food plant quality can cause overexploitation of the food supply by improving larval herbivore survival. This hypothesis was tested and confirmed by Myers and Post (1981) with the cinnabar moth (*Tyria jacobea*), imported to North America as a plant control agent. Nutrient enrichments increased the amount of protein in its host plant, the tansy ragwort (*Senecio jacobea*), and led to both overexploitation by the moth and unstable moth populations. In a similar experiment, Brunsting and Heil (1985) found that an increase in the nutrient content of leaves of *Calluna vulgaris* in heathland through nitrogen fertilization resulted in better growth of the larvae of the monophagous herbivorous beetle *Lochmea suturalis*, so that more severe and frequent outbreaks of the herbivore would be possible. Both of these experiments confirmed Rosenzweig's (1971) paradox of enrichment. Enrichment removes the nutrient constraints that tend to stabilize the systems.

Enrichment may also destabilize ecosystems by increasing the availability of plants to herbivores. Coley *et al.* (1985) proposed that resource availability is the major determinant of plant defence against herbivores: low resource availability results in slower plant growth, greater anti-herbivore defence, and more stable communities. Conversely, it has been suggested that enrichment may result in nutritional imbalances of grasses for ungulates, thus leading to herbivore toxicity and system instability (Ruess, 1984).

The way in which food quality might lead to instability can be seen by examining a simple extension of the cell quota model to include herbivores. In this example, X is supposed to represent the biomass of a large plant and Q is the nutrient concentration of its edible matter. Favourable nutrient availability can raise nutrient concentrations within the cell before the total biomass begins to increase appreciably. The model is

$$dN/dt = I_n - r_n N - [\gamma r_1 XN/(k_1 + N)] + \gamma d_D QD \tag{5.17a}$$

$$dQ/dt = [r_1 N/(k_1 + N)] - \mu_m(Q - k_Q) \tag{5.17b}$$

$$dX/dt = \mu_m(1 - k_Q/Q)X - (d_1 + e_1)X - fX^2Y/(b_1 + X^2) - g_1X^2 \tag{5.17c}$$

$$dY/dt = [\eta f X^2 Y/(b_1 + X^2)] - (d_2 + e_2)Y \tag{5.17d}$$

$$dD/dt = d_2 Y + d_1 X + g_1 X^2 - (d_D + e_D)D + (1 - \eta)fX^2Y/(b_1 + X^2) \tag{5.17e}$$

The positive effect of nutrient concentration will be taken into account by assuming that the mortality rate coefficients of the herbivores ($d_2 + e_2$), falls off as Q increases (Figure 5.7). Can this functional dependence destabilize the system? Rather than carry out a quantitative model study, here I will simply obtain a hint by looking at qualitative feedback diagrams for the system with and without dependence of mortality on Q (Figure 5.8). In Figure 5.8(b), where the extra link representing the positive effect of Q on the herbivore, $Q \rightarrow Y$, is included, the existence of a new positive feedback loop ($Q \rightarrow Y \rightarrow X \rightarrow N \rightarrow Q$), denoted by dashes in Figure 5.8(b), is observed

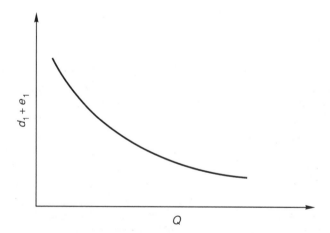

Figure 5.7 Assumed decline in the mortality rate coefficient of herbivores in Equation (5.17d) as internal nutrient concentration, Q, of the autotroph increases.

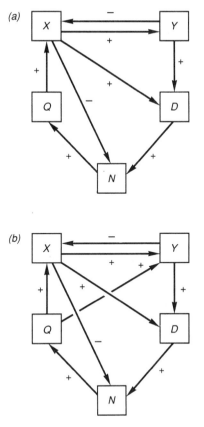

Figure 5.8 Feedback diagram of the system of Equations (5.17), assuming (*a*) that there is no dependence of herbivore mortality on autotroph nutrient concentration, *Q*, and (*b*) that herbivore mortality varies inversely as *Q*.

(i.e. the product of the signs of the links making up the circuit is positive). That particular feedback loop means that an increase in herbivores can suppress autotroph biomass, *X*, taking the autotroph pressure off of the available nutrients, *N*, so that these can build up and increase the cell quota, *Q*, which has a positive effect on the herbivores, *Y*. The existence of such a loop can often be highly destabilizing in a network unless counteracted by strong negative feedback (DeAngelis *et al.*, 1986).

These qualitative results appear to add corroboration that nutrient limitation may act as a stabilizer on some autotroph–herbivore interactions by constraining growth and that enrichment, by releasing this constraint, may result in destabilization. In complex ecological systems no generalization is universal, however, and there are possible exceptions to the rule that greater nutrient limitation should favour food web stability. The negative effect of nutrient quality on herbivore reproduction and survival was mentioned earlier as a stabilizing factor. However, this mechanism has also been hypothesized to be destabilizing. Pitelka (1964) and Schultz (1964, 1969) related lemming cycles in the Arctic tundra to nutrient limitation (see Figure 5.9 for an example in fluctuations of trappable lemmings near Barrow, Alaska). These authors argued that because there is no reserve of available nutrients in tundra ecosystems and a sizeable fraction of nutrients can be tied up in herbivores and litter, periodic crashes of autotrophs and, subsequently, of the herbivores occur. Some aspects of this 'nutrient-recovery' hypothesis, as it is called, have been questioned in the light of subsequent investigations under the International Biological Programme (Batzli *et al.*, 1980), so the hypothesis and its evaluation by Batzli *et al.* will be considered in some detail.

Lemmings reproduce during the winter in nests at the base of the snowpack. If a particular winter corresponds to a lemming population peak, there is a high rate of grazing on live and standing dead graminoids. According to the nutrient-recovery hypothesis (see Figure 5.10 for a schematic of the hypothesis), the reduction in ground cover caused by heavy grazing leads to a reduction in albedo during the following summer, which causes a deeper thaw of the permafrost and allows plant roots to penetrate deeper. There is also a large release of available nutrients in the form of urine and faeces by the lemmings, which stimulates rapid summer uptake of nutrients by plants and high plant nutritive quality. However, later in the summer, nutrient concentration in plants decreases because the plant roots are taking up nutrients from less nutrient-rich lower soil layers. Because of

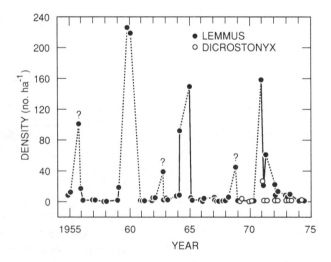

Figure 5.9 Estimates of lemming densities (the brown lemming, *Lemmus sibericus*, and the collared lemming, *Dicrostonyx torquatus*) in the coastal tundra at Barrow, Alaska. (From Batzli *et al.*, 1980.)

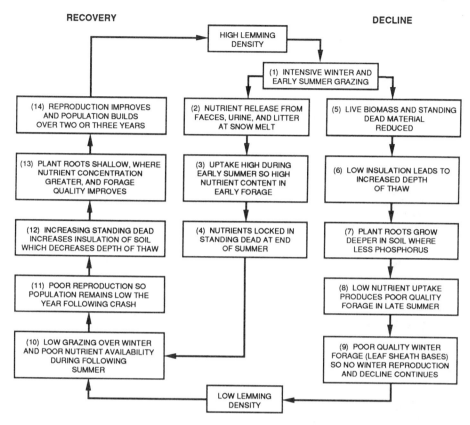

Figure 5.10 Schematic of linkages between causal mechanisms in the nutrient-recovery hypothesis. (From Batzli *et al.*, 1980, adapted from Schultz, 1964, 1969.)

this low quality forage, lemming reproduction during the winter following the winter of their population peak is low and the lemming population declines. This causes a reduction in nutrient regeneration by lemmings the following spring and summer, so that plant nutrient quality remains low and the lemming population continues to decline. According to the hypothesis, it takes a few years for the litter to build up, increase the insulation of the soil, and decrease the depth of the rooting zone, so that plant biomass increases in nutrient concentration. At this point, the lemming population can begin to build back towards a peak.

Batzli *et al.* (1980) confirmed some of the causal mechanisms proposed by the nutrient-recovery hypothesis. Lemming grazing during population peaks can reduce plant biomass substantially (almost 50% of net aboveground production was shown to be consumed). Large amounts of nutrients were returned to the soil surface in faeces and urine, stimulating high early summer plant growth. Other aspects of the hypothesis were not corroborated, however. Increased plant rooting depth appeared to increase, rather than decrease, nutrient availability and plant production (Bliss and Wein, 1972; Challinor and Gersper, 1975; Chapin and Van Cleve, 1978). Batzli *et al.* (1980) proposed that, instead, a somewhat similar mechanism may have resulted in lower plant nutritional quality. During their population peaks, lemmings grazed intensively and grubbed for rhizomes to the extent that plant phosphorus reserves were depleted and required time to recover. Weasel and other predation may have also played a greater role in the lemming crashes than the original hypothesis assumed.

Whatever the exact mechanisms may be, the lemming cycles may represent a nutrient-related instability that is different from the paradox of enrichment. It may arise from an inherent time lag in the system, a factor that has not been considered here before. The reduction in available nutrient for the plants requires a period of a few years to be restored, during which time the lemming populations can fall to extremely low numbers relative to their peaks. Time lags of this nature are frequently implicated in population oscillations (e.g. see May, 1973). Schultz (1969) suggested that nutrient enrichment of the soil by artificial fertilization could eliminate the time lag and thus eliminate that mode of instability. He performed fertilization experiments, which improved the quality (protein, phosphorus and calcium levels) in fertilized graminoids. The reproductive success of lemmings in the fertilized areas was greatly improved over controls in the following winter (75 nests ha^{-1} versus none).

The possibility of instability induced by nutrient limitation in combination with time lags also may exist in aquatic ecosystems. In arctic and temperate regions, spring blooms of phytoplankton may deplete surface nutrients before the herbivorous zooplankton can respond. Also, Cushing (1961) hypothesized that the Plymouth herring fishery failed in some years because the limiting nutrient, phosphorus, was locked up in the biomass of pilchards (another herring-like fish) rather than being available to the plankton prey of larval herring.

A further feasible destabilizing effect of low nutrient availability was noted by Moran and Hamilton (1980). Low nutrient levels in foliage could cause certain herbivores to consume more plant material to balance their nutrient needs. When the effect of low nutrient levels in the autotrophs is not to cause the herbivores to die but to cause them to feed more voraciously to acquire sufficient nutrients, this is a depensatory effect; that is, the pressure on resources becomes greater the smaller the resource pool becomes. This depensatory effect is potentially destabilizing, at least for short time periods. As an example, Barkley *et al.* (1980), investigating the nutrition of microtine rodents, found that the nutrient levels of graminoids can change from year to year. Using a simulation model of lemming growth and reproduction, the authors showed that lemmings feeding on mosses could not always get enough nutrients in the process of feeding to satisfy energy demands, and thus had to feed to excess. Low nutrient concentrations during some years produced severe depletions in adult females, which resulted in low reproduction. Thus, over longer time scales, the effect of low nutrient availability is to lower rodent population levels, lessening the rate of herbivory.

A number of studies of mammal herbivores have indicated that their feeding strategies may reflect limitation by sodium. For example, Belovsky and Jordan (1981) noted that moose on Isle Royale feed on aquatic plants in midsummer, which is a poor strategy from an energetic point of view, but which may alleviate sodium stress.

5.5   EFFECTS OF NUTRIENTS AND HERBIVORES ON RESILIENCE

The third major property of food web models that we consider here is resilience, which was defined in Chapter 2 to be the rate of return to steady state following a perturbation or the inverse of the return time T_R following a perturbation. The return time to steady state was shown to be similar to the turnover time, T_{res}, of nutrient between the model system and the external world. In Chapter 2 the resilience of a simple flow-through nutrient pool was examined, and in Chapter 3 I computed the resilience of a nutrient pool–autotroph–detritus system. Here we extend the analysis of resilience to a system that also includes an herbivore.

In Chapter 3, in discussing resilience, I imagined a forest ecosystem in a region of very low external nutrient input. The trees were highly effective at recovering available nutrients from the soil and litter. Thus, the tree biomass per unit area was large in this imaginary system and the turnover of nutrients between the system and the external environment was low; available nutrient in the soil water was low so that loss rates due to export in waterflow was low. It was shown that such a system has low resilience to disturbances that remove large amounts of biomass, because the ratio of new nutrient input to standing stock of nutrients is low. Now suppose, however, that there is a high level of herbivory in the system. This would cause a faster return of nutrient from the autotroph to the available nutrient pool in soil water, raising soil water nutrient levels and increasing the loss rates. Initially, the effect is to decrease the biomass of autotrophs and to decrease the turnover time of nutrients between the forest system and the external world. Because resilience of a system is inversely related to turnover time, it is expected that herbivory may increase the resilience of the forest system. This intuitive discussion will be tested by model analysis.

The equations used are simply those of Equations (5.16a,b,c,d) with $F = fXY/(b + X)$ replaced by $F = fX^2Y/(b_1 + X^2)$ to confer local stability on the system. The specific parameter values used in this case are shown in Table 5.1, which is an extension of the set of values of Table 3.3. There is nothing very special about this set of parameter values, although the herbivore biomass Y has a more rapid turnover time than autotroph biomass X and there is a high degree of recycling. The results for the return

Table 5.1 Parameter values used in Equations (5.16a,b,c,d) to examine resilience characteristics. No specific units are intended for these values and do not apply to a particular system

I_n = 0.00005 to 0.002	f = 2.0	e_1 = 0.001
r_n = 0.005	d_1 = 0.1	e_2 = 0.001
k_1 = 0.005	d_2 = 0.1	e_D = 0.001
k_2 = 60.0	d_D = 0.1	η = 0.5
γ = 0.02		

time as a function of I_n, calculated from simulations and the use of Equation (2.8), are shown in Figure 5.11 (solid dots and line). The return time falls off rapidly as a function of I_n over a large range of values of I_n. This is quite different from the case of the autotroph alone (Figure 3.8, shown again here by the open dots in Figure 5.11). What is most interesting is that for large values of I_n the return time for the present system is much smaller than for the case where only the autotroph was present.

As in the preceding case where only the autotroph was present, we can

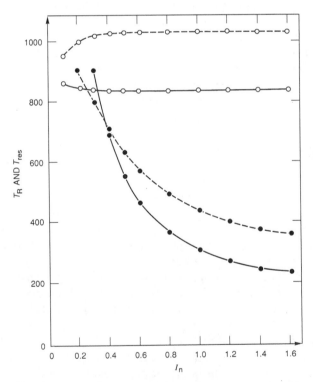

Figure 5.11   Return times, T_R, of system following a perturbation for four cases: (1) Autotroph alone (open dots, solid line); (2) autotrophs and herbivores (closed dots, solid line). The dashed lines represent corresponding nutrient turnover times, T_{res}, in the two situations. Nutrient input, I_n, is multiplied by 10^3 in the figure.

try to gain some insight into the behaviour of the system by performing analysis of the equations. The autotroph is held at a fixed level, X^*, by the herbivore, where, from Equation (5.16c), with Holling Type II replaced by Holling Type III functions:

$$X^* = [b_1(d_2 + e_2)/(\eta f_1 - d_2 - e_2)]^{1/2} \qquad (5.18)$$

whereas N^* is a monotonically increasing functions of I_n. The expression for N^* is complex for small values of I_n, but for very large values of I_n, a great simplification occurs and N^* has the form

$$N^* \to I_n/r_n \quad \text{as} \quad I_n \to \infty$$

There is a threshold value of I_n of about 0.15×10^{-3}, below which the herbivore cannot survive in the system. Plots of N^*, X^* and Y^* as functions of I_n are shown in Figure 5.12. From this plot it is seen that the present case

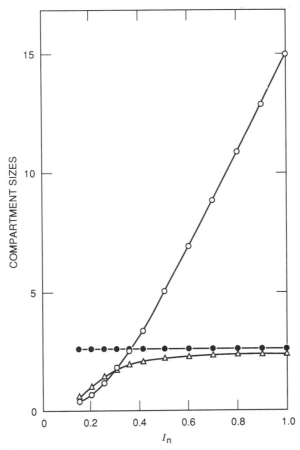

Figure 5.12 Steady-state values of available nutrient, N^*, multiplied by 10^2 (open circles), autotroph biomass, X^* (closed circles) and herbivore biomass, Y^* (triangles) as functions of nutrient input, I_n, multiplied by 10^3.

represents a reversal in some sense of the case in Chapter 3, in which the autotroph held N^* fixed while X^* increased linearly with I_n. Here X^* is held fixed from above and N^* can increase with I_n. It is not surprising that the resilience characteristics of the two systems differ.

When the turnover time of nutrients in the system, $T_{res} = N_T^*/I_n$ (where $N_T^* = N^* + \gamma X^* + \gamma Y^* + \gamma D^*$) is plotted as a function of I_n, this ratio falls off rapidly to an asymptote with increasing I_n (solid dots and dotted line in Figure 5.11). Thus, the mean nutrient turnover time, T_{res}, agrees reasonably well with the computed return time to steady state, T_R, in the case where there is an herbivore, as T_{res} and T_R agree in the case where only the autotroph is present (open circles in Figure 5.11).

The reason for the pronounced difference from the case where the autotroph is the top trophic level follows from the three facts: (1) the autotroph biomass is limited by the herbivore, (2) the herbivore compartment size also reaches an upper limit (as does the detritus compartment) and (3) the available nutrient N^* is only a small fraction of total nutrient in the system for small values of I_n. Because of these three properties N_T^* initially increases only very slowly with increasing I_n. Thus, N_T^*/I_n falls off as shown by the closed circles and dashed line in Figure 5.11. Only when I_n is large enough that the level of available nutrient N^* becomes significant, does N_T^* increase linearly with I_n, so that the curves in Figure 5.11 flatten out.

For small values of I_n, the computed values of T_R increase more rapidly with decreasing I_n than does N_T^*/I_n. This occurs because for small values of I_n the recycling of nutrients is not as tight as when I_n is large. In particular, it can be shown that a relatively smaller proportion of the nutrient input I_n passes through the herbivore compartment as I_n decreases. Consequently, the return time of the system increases rapidly as I_n decreases to low values.

In summary, the ratio N_T^*/I_n provides a reasonable description of the dependence of the system return time T_R on I_n, except for differences that occur at low values of I_n, where nutrient cycling is not tight. The short return time of the system with a herbivore as top trophic level is explicable in the way the behaviour of N_T^* as a function of I_n is affected by the trophic structure.

5.6 SUMMARY AND CONCLUSIONS

This chapter has reviewed one general aspect of the interactions of herbivory and nutrient cycling in ecosystems, namely the effects that nutrient limitation and cycling have on autotroph–herbivore interactions. These interactions are a special case of consumer–resource interactions, which have been a major object of study in ecology.

Nutrient limitation has a direct effect on the steady state of an ecological system, in particular on the number of trophic levels that can be supported. A threshold level of available limiting nutrient is needed for autotrophs to become established and a higher level is needed to raise autotroph biomass to the level where herbivores can invade and become established. These thresholds can be calculated for simple models.

One of the most interesting results related to nutrient effects is the paradox of enrichment, a model prediction that increasing the level of available nutrient in an autotroph–herbivore system may lead to instability. This prediction seems to be a robust one, as a diverse array of models predicts its occurrence. A large body of experimental research has by now shown the existence of nutrient-induced oscillations. Apparently, many autotroph–herbivore relationships become unstable when the growth of the autotrophs is not sufficiently constrained by nutrient limitation.

For autotroph–herbivore systems that are locally stable (that is, that return to steady state following perturbations) nutrient limitation may still play a role in the system resilience, or rate of return to steady state following a perturbation. It is interesting that a nutrient–autotroph model may sometimes be less resilient than the same model to which an herbivore has been added. This may be the case when the presence of the herbivore causes the total biomass of the system to be less than in the case where the herbivore was not present. The result of this is that the nutrient turnover time in the system with the herbivore is less than it is in the system without the herbivore. Since resilience (the inverse of return time) is inversely related to turnover time, the system with the herbivore is more resilient. This will be discussed further in Chapter 8.

6 Herbivores and nutrient recycling

6.1 INTRODUCTION

Two important aspects of how nutrient limitation affects an autotroph–herbivore system were discussed in the preceding chapter: the influence of nutrient input levels on the steady-state values and the influence on system stability. In a sense, the present chapter is complementary to the concerns of Chapter 5. Whereas that chapter was largely concerned with how nutrient limitation alters the stability and steady-state values of an autotroph–herbivore system, Chapter 6 is primarily concerned with the complementary problem of the indirect effects of the herbivores, through nutrient recycling, on the available nutrients and autotrophs of an ecosystem.

The main topic of this chapter is the question of what effect enhanced recycling due to the presence of herbivores can have on living biomass and productivity in an ecosystem context. There are some interesting questions within this topic that will receive particular attention, including whether herbivory can have a net benefit due to a faster rate of recycling of nutrients.

6.2 EFFECTS OF HERBIVORE-ENHANCED NUTRIENT CYCLING ON AUTOTROPHIC PRODUCTION

The study of herbivore–autotroph interactions historically has focused almost exclusively on the gains in energy to the herbivore and the losses of biomass from the autotrophs that occurred. These energetic gains and losses were assumed to be the essential features of the interactions as they affected the dynamics of the two populations. This view is changing, however. In phytoplankton research, for example, it has been recognized for a few decades now that phytoplankton cells use nutrients at far greater rates than can be accounted for by external inputs of inorganic nutrients (e.g. Barlow and Bishop, 1965; Lehman, 1978, 1980). High uptake rates of ammonium and urea have also been noted in phytoplankton (Eppley et al., 1973) suggesting that excretion from zooplankton is a source of nitrogen for the phytoplankton. This evidence and evidence of the effects of biogeochemical recycling, facilitated by herbivores, between terrestrial plants and their environments (see below) have led in recent years to a reassessment of the traditional assumption that herbivory is always deleterious to autotrophs, as reflected by the following titles: 'How plants may benefit from the animals that eat them' (Owen, 1980), 'Do consumers maximize plant fitness? (Stenseth, 1978), 'Enhancement of algal growth and productivity by grazing zooplankton' (Porter, 1976).

Kitchell et al. (1979) described many ways in which consumers affect the recycling of nutrients. For example, in terrestrial systems the effect of consumption by herbivores is to increase nutrient cycling rates by removal of senescent individuals and plant parts. Also, nutrient pools stored in large biomass units may be transformed into smaller units by consumption, promoting accelerated nutrient cycling processes. In addition, there are processes of translocation due to migration of fish and zooplankton that increase the uniformity of nutrients in space. Although these and a variety of other mechanisms have been suggested by which herbivory might benefit autotrophs, the proposal of interest here is that consumers may provide help by facilitating the cycling of a limiting nutrient (Owen, 1980). At least superficially, this seems feasible because plant productivity is more often restricted by the supply of nutrients than by energy input (Witkamp, 1971). Others supporting this idea include Rafes (1970), Reichle et al. (1973) and Kimmins (1972).

The growing interest in possible herbivore benefits to autotrophs has developed in parallel for aquatic and terrestrial systems. Porter (1976) reported this phenomenon for certain species of planktonic green algae, Sphaerocystis schroeteri, which normally pass through the gut of the predator Daphnia magna relatively unscathed. In fact, the increased growth due to enhanced uptake of nutrients such as phosphorus within the gut may more than make up for occasional losses of individuals from the population due to Daphnia grazing. It is possible that similar small-scale concentrations of nutrients in the vicinity of zooplankton have a large effect on phytoplankton growth under oligotrophic conditions. Some plankton studies have shown that zooplankton excretion could supply a large part (perhaps all) of the nutrient needs of phytoplankton during periods of phytoplankton scarcity (Martin, 1967). In Lake Michigan, indirect interactions between the zooplankter Mysis relicta and other organisms, such as small zooplankton and fishes, could be major factors in nutrient recycling within the metalimnion and subthermocline region (Madeira et al., 1982).

The positive effect of nutrient recycling by grazing Daphnia has been demonstrated even in the case of algae that are less resistant to mortality through Daphnia consumption than are the green algae (Sterner, 1986). Sterner's experimental procedure was to establish a gradient in herbivore density and to monitor phytoplankton growth rates in the presence of both Daphnia pulex grazing and nutrient regeneration and Daphnia nutrient regeneration alone, the latter being made possible by separating some of the phytoplankton from Daphnia by exclosures that were porous to water and nutrients. Sterner plotted the net growth of algae per day in the experimen-

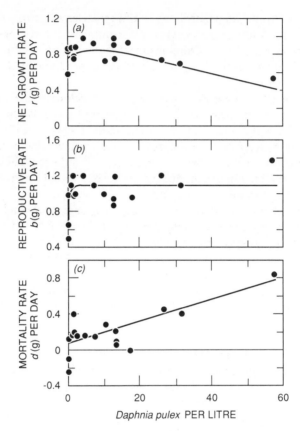

Figure 6.1 Net growth of algae per day as a function of *Daphnia* density in system, representing possible positive effects of nutrient recycling on algal growth. (From Sterner, 1986).

tal system versus the number of Daphnia per litre, as shown in Figure 6.1. Figure 6.1(*b*) shows the indirect effect of nutrient regeneration on phytoplankton reproductive rate inside *Daphnia* exclosures, while Figure 6.1(*c*) shows the effect of *Daphnia* on phytoplankton mortality alone (the negative values of mortality are due to sampling errors). The net growth is the difference between the reproductive rate per day and the mortality rate per day of algae. The two effects combined yield the curve in Figure 6.1(*a*). Note that there is a peak for moderate densities of *Daphnia* in the system, indicating that the algae grow best when subjected to some grazing.

The possibility has also been raised that a change in the size structure of zooplankton could affect the nutrient cycling rate and thus autotroph production rates. The size distribution of zooplankton present in a body of water affects the phytoplankton size distribution because of differential grazing by zooplankton of different sizes (Porter, 1977; Steele and Frost,

1977; Gliwicz, 1977, 1980; DeMott, 1982). This effect on phytoplankton size structure can affect primary production as a whole, because smaller phytoplankters have higher maximum growth rates at low nutrient levels than do large phytoplankters (Malone, 1980).

Carpenter and Kitchell (1984) combined all of these factors into a single model. They modelled nutrient-limited phytoplankton, which they divided into ten species populations; each with a different characteristic size. The growth rate of each species was described by a Monod-type growth function of nutrient concentration N, in which the half-saturation constants, k_1, increased linearly with phytoplankter cell radius. The rate of algal loss from a given size class depended on the mean zooplankton biomass, Y (mg dry matter l^{-1}) and mean individual zooplankton size, S_y (mg dry matter animal^{-1}). Feeding was size dependent. A zooplankter of a given size, S_y, could only consume phytoplankton of sizes smaller than some radius (determined elsewhere from feeding experiments; e.g. Gliwicz, 1980; Richman *et al.*, 1980). In addition to nutrient input, I_n, to the system, the model allowed for regeneration of nutrients by zooplankton that was inversely proportional to mean zooplankton size (Peters and Rigler, 1973; Peters, 1975; Hall *et al.*, 1976; Bartell and Kitchell, 1978).

Carpenter and Kitchell (1984) determined the parameters for their model from physiological data on zooplankton and phytoplankton and used the model to examine primary production as a function of given values of $\log(Y)$ and $\log(S_y)$ (Figure 6.2). The figure shows a striking ridge-like

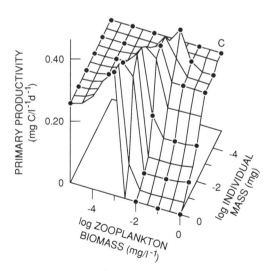

Figure 6.2   Response surface of primary production to zooplankton biomass, Y, and mean individual zooplankter mass, S_y, (both as logarithms to the base 10) from model simulations performed by Carpenter and Kitchell (1984). (From Carpenter and Kitchell, 1984.)

unimodal response, peaked at intermediate zooplankton biomass densities, that was relatively insensitive to changes in parameter values. This phenomenon is explainable as follows. At low grazing pressure, larger phytoplankters outcompete the more vulnerable smaller phytoplankters, so that primary production was dominated by the relatively low growth rates of the larger-sized phytoplankters. At very high rates of herbivory, primary production is low becuase the phytoplankton biomass density was depressed. At intermediate rates of herbivory, the combination of moderately high phytoplankton survival, relatively small phytoplankton size (with high specific growth rates) and high nutrient recycling by zooplankton maximized primary production. Note in Figure 6.2 that there is also an increase in primary production along the ridge as zooplankter size, S_y, decreases, which may reflect the positive effect of greater nutrient recycling by smaller zooplankters.

Carpenter and Kitchell (1984) cited several examples of empirical evidence that seems to corroborate the model results. Cooper (1973) noted from aquarium experiments that filamentous algae had their highest rates of production at intermediate densities of a herbivorous fish. In a field study, Seale (1980) reported the highest specific production rates of phytoplankton at intermediate levels of herbivorous tadpoles.

It is interesting to go one step further in the trophic chain and explore an additional causal link, that of planktivorous fish, on nutrient recycling. Bartell and Kitchell (1978) used a combination of zooplankton sampling in a temperate lake (Lake Wingra, Wisconsin) and statistical analysis to predict the effects of planktivory by bluegills (the major fish species in the lake) on the zooplankton standing stock and size distribution and, indirectly, on nutrient recycling. Bluegill feeding rate varies during the year due to seasonal changes in temperature and bluegill age structure. Thus, changes in the level of effect of planktivory on the zooplankton occur. Because fish predation on zooplankton is generally size-selective for the larger zooplankton sizes (e.g. Brooks and Dodson, 1965), high levels of bluegill predation should alter the zooplankton size distribution towards smaller size classes. One implication of such a size distribution shift would be a higher excretion rate of nutrients per unit zooplankton biomass, since turnover rate is higher in smaller zooplankters.

Bartell and Kitchell (1978) measured the standing stock of zooplankton through the year in the presence of bluegills and used estimates of bluegill feeding rates and size selection of zooplankton to estimate the zooplankton abundance and size distributions in the absence of bluegills. Using a regression model for phosphorus release from zooplankton as a function of zooplankton size (Peters and Rigler, 1973), they computed what the indirect effects of planktivory would be on P release through the year. Bluegills were estimated to have a maximum effect on zooplankton during the months of June and July, when planktivory is highest. Both the standing stock and size distribution should be significantly affected during these months. Although a decrease in the mean size of zooplankton was predicted, which should increase the per capita release of P, the depression of zooplankton standing stock during this time overwhelmed any effects of size distribution and lowered the release of P by about an order of magnitude during these months.

Bartell and Kitchell noted that the release of P by fish cannot compensate for this drop in P release by phytoplankton during June and July. Direct remineralization of P by fish occurs at rates several orders of magnitude slower in fish than in zooplankton (Seadler and Koonce, 1976). Since fish biomass may tie up to 70% of pelagic P in Lake Wingra during the summer, fish can be a major sink for the nutrient most limiting to primary production.

In terrestrial systems results similar to the effects of herbivory in aquatic systems described above have been summarized in the 'herbivore optimization curve' (HOC) hypothesis (McNaughton, 1979a). The HOC hypothesis states that productivity of plants increases with light grazing, then drops off and becomes negative as grazing becomes severe (Dyer *et al.*, 1982; Hilbert *et al.*, 1979) (Figure 6.3). Thus far, evidence for the HOC has been found in many plant–animal associations. Table 6.1 lists a number of examples from a variety of studies in different systems. These studies are highly diverse and it is possible that many different factors were involved in producing the observed HOCs. Also, because it is difficult to measure all biomass production, especially belowground production, it is possible that in some cases measurements were not thorough enough to determine whether total growth had actually increased (Verkaar, 1986).

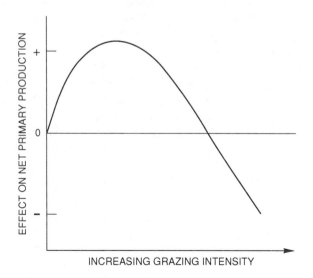

Figure 6.3 Hypothesized 'herbivore optimization curve' indicating the maximization of productivity of plants under light grazing. (From McNaughton, 1979a.)

Table 6.1 Examples from terrestrial systems of the herbivore optimization curve (HOC) as cited in Dyer *et al.* (1986)

Plant type	Herbivore	Reference
Grasses and sedges	Native and domestic grazers	McNaughton (1979a,b, 1984), Vickery (1972), Coppock *et al.* (1983a,b), Williamson (1983)
Grasses	Small mammals	Hilbert *et al.* (1979)
Grasses	Simulated herbivory in laboratory	Danell *et al.* (1985), Detling and Dyer (1981), Detling *et al.* (1979), Dyer (1975), McNaughton *et al.* (1983)
Plantain	Geese	Drent (1980), Prins *et al.* (1980)
Hybrid field corn	Birds	Dyer (1975)
Wheat seedlings	Insects	Capinera and Roltsch (1980)
Flower production in higher plants	Caterpillars	Hendrix (1984), Solomon (1983)
Juvenile birch	Moose	Danell *et al.* (1985)
Tropical successional communities	Simulated grazing	McNaughton (1984)

However, painstakingly careful studies continue to show similar results (Paige and Whitham, 1987). These authors studied the effects of mammalian herbivory on scarlet gibia, *Ipomopsis aggregata*, in northern Arizona. Plants that were browsed (up to 95% of aboveground biomass) were found to overcompensate in the numbers of inflorescences, flowers and fruits produced, as well as in total plant biomass.

In terrestrial systems, as in aquatic systems, one possible explanation for the observed HOC curve is that the consumption by grazers may stimulate nutrient cycling processes and may, therefore, benefit plant growth by increased availability of such elements as nitrogen, phosphorus and potassium (Rafes, 1970; Owen, 1980; Schowalter *et al.*, 1981). Seastedt *et al.* (1983) measured an increase in the recycling rate of K due to herbivory, though no net effect on biomass production of black locust and red maple trees in the southern Appalachian Mountains was observed.

6.3 MODELS OF THE EFFECTS OF HERBIVORE NUTRIENT RECYCLING ON AUTOTROPHIC PRODUCTION

Further insights concerning the effects of enhanced nutrient recycling on biomass standing stocks and production rates can be obtained by detailed investigation of mathematical models. This is especially useful because intuition alone is often a poor guide to determining the effects of complex feedback processes. Consider first the model represented by Equations 5.11. In the simulations shown in Figure 5.5, none of the limiting nutrient was recycled. Suppose, however, recycling is incorporated by reducing the loss rates e_1 and e_2 from the system (from $e_1 = 0.20$ to 0.05 and $e_2 = 1.01$ to 0.01) and correspondingly increasing the rates, d_1 and d_2 (from $d_1 = 0.0$ to 0.15 and from $d_2 = 0.0$ to 1.0), at which nutrients from the autotrophs (phytoplankton) and herbivores (zooplankton) are channelled through the detritus and released back to the available nutrient pool, such that there is a recycling index [*sensu* Finn, see Equation (3.16)], $CI = 0.70$. The results of the revised computations of steady-state biomass are shown in Figure 6.4. Note that when the herbivore is present in the system, it regulates the autotroph cell density to the same level; $X^* = (d_2 + e_2)/(\eta f) = 1.01/(0.5 \times 10^{-7}) = 2.02 \times 10^{-7}$ g. Note, however, that the strong recycling stimulates a higher nutrient level and higher zooplankton biomass in the second case. The rate of phytoplankton production is increased, therefore, because steady-state levels of available nutrient, N^*, are increased by recycling.

This simple result indicates that production rates and standing crops are increased when recycling is increased. There are also other approaches to questions of the relationship of recycling and production. For example, is recycling adjusted by the species populations involved in some way to optimize energy usage?

Kiefer and Atkinson (1984) used the idea of optimal energy usage in attempting to predict how a limiting nutrient, nitrogen in their case, would be divided among the components of a marine euphotic zone ecosystem.

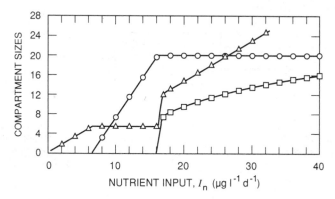

Figure 6.4   Changes in the nutrient concentration (triangles) in water, N (μg l^{-1}), number of phytoplankton cells ($\times 10^7$ l^{-1}) (circles) and herbivore (zooplankton) cells (10^6 l^{-1}) (squares) as functions of nutrient input I_n. This is the same as Figure 5.5, but with recycling of limiting nutrient (recycling index $CI = 0.70$).

These components were standing stock of NH_4^+ dissolved in water in the system termed N here, standing stock of nitrogen stored in phytoplankton, N_x, and standing stock of nitrogen stored in zooplankton nutrient concentration, N_y, all in units such as $g\,m^{-3}$. In the notation of this book, the latter two variables would be $\gamma_x X$, and $\gamma_y Y$, where γ_x and γ_y are proportions of the limiting nutrient in the autotroph and herbivore biomasses respectively. For simplicity here, however, Keifer and Atkinson's symbols will be used. The system, pictured in Figure 6.5, was assumed closed with respect to nutrients; that is, there were no inputs or outputs of nitrogen, but only fluxes, F_i, between compartments.

Kiefer and Atkinson (1984) characterized their system using a minimum amount of known information. The overall nature of the system was described by three parameters: Z_m, the depth of the upper mixed euphotic zone; $E_0(0)$, the photosynthetically active irradiance immediately below the sea surface; and N_T, the total concentration of nitrogen in the system. Two known relations characterized the constraints of the system:

1. Conservation of nutrient in euphotic zone

$$N_T = N + N_x + N_y \tag{6.1}$$

2. Steady-state nutrient flux balance through phytoplankton

$$F_2 = F_3 + F_4$$

or

$$f_2 N_x = f_3 N_x + f_4 N_y \tag{6.2}$$

where

f_2 = specific rate of ammonium assimilation
$\quad = f\,[E_0(0),Z,N,CHl]$ $\qquad\qquad\qquad$ (6.3)

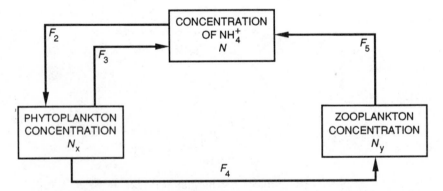

Figure 6.5 Schematic of a nitrogen cycling system of the mixed upper ocean layer. The system is assumed to be closed and nutrient is recycled by zooplankton as ammonium to the available pool. (Adapted from Kiefer and Atkinson, 1984.)

f_3 = specific rate of ammonia loss to respiration

f_4 = specific rate of nitrogen ingestion by zooplankton

$\quad = f(N_x)$ (6.4)

and where CHl is the chlorophyll concentration and Z is depth.

The functional forms of f_2 and f_4 were specified by Kiefer and Atkinson in considerable detail, using data for the phytoplankton species *Chlamydomonas reinhardi* and the zooplankton species *Daphnia pulex*. Liebig's Law of the Minimum was used to specify phytoplankton growth limitation either by the nutrient (ammonium) or by light. In either case a Monod-type growth function was assumed, where Beer's Law of Light Extinction was integrated into the light limitation function. The specific rate of nitrogen ingestion by zooplankton depended on individual zooplankton filtration rates, the numerical concentration of phytoplankton and the amount of nitrogen per individual phytoplankter, as adapted from a model by Lehman (1976). The greatest efficiency of conversion of phytoplankton nitrogen to zooplankton nitrogen occurs at intermediate levels of phytoplankton concentration. At these levels, the zooplankton are well fed enough that they are not suffering net energy losses through basal metabolism but are not oversaturated with phytoplankton numbers to the extent that the transfer from phytoplankton to zooplankton biomass is inefficient.

Equations (6.1) and (6.2), supplemented by the functional forms (6.3) and (6.4), are not sufficient to determine the actual amounts of standing stock in the three compartments, N, N_x and N_y. One way to make this determination is, for instance, to specify a particular rate of urea production by the zooplankton. However, Kiefer and Atkinson took another approach. They assumed that the system might be regulated in such a way that the standing stocks maximize the fraction of incident radiant energy that is converted to zooplankton biomass. A quantity η_T was defined as the 'efficiency of radiant energy conversion to zooplankton biomass' or the 'system efficiency'. This quantity is a function of zooplankton and phytoplankton number concentrations, zooplankton filtration rate, energy content of phytoplankton cells, cell basal metabolism and solar irradiance. The authors used a computer algorithm for maximization of η_T, in conjunction with Equations (6.1, 6.2, 6.3, 6.4), to calculate the standing stocks N, N_x and N_y. In particular, they iterated over phytoplankton cell number concentration to find the maximum value of η_T.

The results of maximizing η_T as a function of phytoplankton cell number (in a litre of water) are shown in Figure 6.6. This plot shows not only η_T (system efficiency), but also the growth efficiency of the zooplankton (calories of new tissue per calorie of ingested phytoplankton) and phytoplankton efficiency of radiant energy conversion. Note that the highest system level efficiency occurs at the boundary between light and nutrient limitation of the phytoplankton.

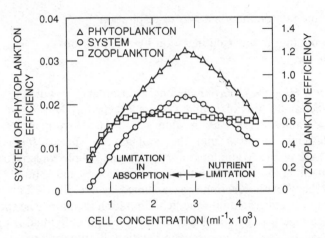

Figure 6.6 Efficiencies of conversion for fixed upper ocean layer model: efficiency of conversion of radiant energy to phytoplankton, efficiency for conversion of phytoplankton biomass to zooplankton biomass and total efficiency of conversion of radiant energy to zooplankton biomass. (From Kiefer and Atkinson, 1984.)

Given this determination of nutrient standing stocks in ammonia, phytoplankton and zooplankton, one can further ask what effects various factors, such as total nutrient level in the system, N_T, maximum solar irradiance, E_0, and depth, Z, have on the relative compartments. Comparisons with field data made by Kiefer and Atkinson show that the concentrations of nutrient stocks in the various compartments as functions of N_T, E_0 and Z in the model corresponding to maximum system efficiency are in good agreement with field data. One especially interesting conclusion is worth mentioning here. When the nutrient stocks in different compartments are plotted against total nutrient, N_T, in the system (Figure 6.7), both phytoplankton and zooplankton are seen to be limited at very low levels of total nitrogen. As Kiefer and Atkinson point out, this result is of interest because it may cast doubt on suggestions (McCarthy and Goldman, 1979) that ammonium recycling in oligotrophic oceans should prevent phytoplankton from ever being nitrogen limited.

The model by Kiefer and Atkinson indicates that a population of zooplankton, through their control over the rate of recycling of nutrients, could in principle maximize the efficiency of conversion of radiant energy into their own biomass. It is interesting that this maximization is achieved largely through increasing the efficiency of phytoplankton production. This indicates that, if herbivory occurs according to some scheme that optimizes herbivore benefits, it may also in some sense benefit the grazed autotroph population. Of course, it may not benefit individual autotrophs that are consumed.

Large autotrophs such as grass, shrubs and trees can be grazed or

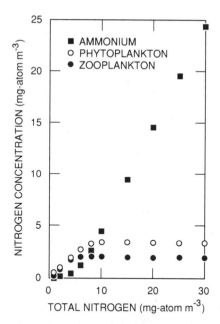

Figure 6.7   Standing stocks of nitrogen in the three compartments of the model of the upper ocean layer as a function of total nitrogen in the system, N_T. (From Kiefer and Atkinson, 1984.)

browsed without causing mortality to the individual. The above model, adapted to these large plants, would show that herbivory increases the efficiency of conversion of incident light into plant biomass for individual plants. Along these lines, Stenseth (1978) considered an abstract model of plant growth in which a given level of herbivory simultaneously caused reductions in biomass of the plant and increases in fecundity. The fecundity increased due to higher production efficiency that accompanied higher nutrient levels in the soil (this assumption of higher nutrient levels depended on the herbivores depositing their wastes directly in the vicinity of the plant). Stenseth showed that if the recycling substantially improved reproductive success, herbivory could confer an advantage to the plant, and represent an evolutionarily stable strategy.

To conclude this discussion, it is possible to show in a model of nutrient recycling that herbivory may increase the net growth rate (even when consumed biomass is subtracted) of a nutrient-limited autotroph. For simplicity, let us use the model of the pelagic phytoplankton population (O'Brien, 1974) once more. Consider the total system shown in Figure 6.8, where, as before, N is the pool of available nutrients per unit area, X is phytoplankton cell number, Y is zooplankton cell number and D is the detrital biomass. Assume that the phytoplankton sink from the euphotic zone carrying nutrients at a rate $e_1 X$, that detritus does not recirculate

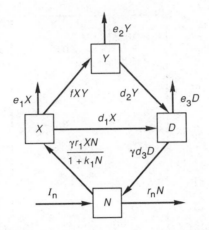

Figure 6.8 Model system consisting of nutrient pool, N, autotroph biomass (or numbers), X, herbivore biomass (or numbers), Y, and detrital biomass, D. The arrows indicate transfers of nutrients or biomass.

nutrients back to the pelagic system, and that zooplankton feed on phytoplankton at a rate fXY and recirculate all of the nutrients at a rate d_2Y. This is an extreme assumption, meant to maximize the nutrient recycling benefits of zooplankton. The equations of the system are

$$dN/dt = I_n - r_n N - [\gamma r_1 XN/(k_1 + N)] + \gamma d_2 Y \tag{6.4a}$$

$$dX/dt = [r_1 XN/(k_1 + N)] - fXY - e_1 X \tag{6.4b}$$

$$dY/dt = fXY - d_2 Y - e_2 Y \tag{6.4c}$$

where the autotrophs have Monod-type nutrient-limited growth and detritus is not explicitly modeled. The term fXY represents the Lotka–Volterra assumption for consumption, and d_2Y is the direct recycling of nutrients back to the nutrient pool. As before, γ is the ratio of nutrient mass per unit biomass in the autotroph and herbivore components are assumed to be identical here. In this model all of the phytoplankton nutrient that ends up in detritus is lost to the system through sinking and all of the phytoplankton nutrient that ends up in the zooplankton is recirculated ($d_2 = 1.0$, $e_2 = 0.0$). It is interesting to consider the steady-state situation ($dN/dt = dX/dt = dY/dt = dD/dt = 0$) and to ask if increases in herbivory can increase the phytoplankton production rate, $r_1 NX/(k_1 + N)$. Intuitively, one would expect that, as herbivory increases, the level of available nutrient in the pelagic system will increase, because an increasing fraction of nutrient is recycled back to the available nutrient pool by the herbivores relative to that lost to detritus. It might be further expected that this would have a stimulating effect on phytoplankton production. To test this idea, one can compute the steady-state values of N, X^* and Y^* from Equation (6.4), which are (for $e_2 = 0$)

$$N^* = (I_n - \gamma d_2 e_1/f)/r_n \tag{6.5a}$$

$$X^* = d_2/f \tag{6.5b}$$

$$Y^* = \{[r_1 N^*/(k_1 + N^*)] - e_1\}/f \tag{6.5c}$$

From the expression for N^* (6.5a), it can be seen that the available nutrient is enhanced by the increase in zooplankton herbivory. However, phytoplankton numbers vary inversely with increasing f. Some combinations of parameter values can be found such that increasing f causes $r_1 N^*/(R_1 + N^*)$ to increase faster than X^* decreases, so that phytoplankton production, $r_1 X^* N^*/(k_1 + N^*)$ may tend to increase with increasing f. However, a systematic study of Equations (6.5) shows this to be an unusual situation.

As an alternative, I assumed that an increase in herbivory, f, might be accompanied by a corresponding decrease in the per capita loss rate of phytoplankton to detritus, e_1. That is, I assumed that zooplankton acted as a competing risk that reduced the risk of phytoplankton cells dying naturally and sinking from the euphotic zone. A number of different assumptions might be hypothesized as to how the two competing risks of phytoplankton mortality would affect each other. In the cases that I considered, which involved simply letting e_1 decrease by some proportion as f increased, such that the increased mortality rate of herbivory on phytoplankton was partially offset by decreased sinking losses, this increased the tendency for phytoplankton production to increase with increasing herbivory. An example is shown in Figure 6.9. In this simulation the phytoplankton production has a peak at intermediate values of consumption rate. Thus, while enhanced recycling by herbivory does not necessarily increase phytoplankton production through recycling of nutrients alone, it can certainly do so if it reduces competing mortality risks that result in a permanent loss of nutrients.

6.4 SUMMARY AND CONCLUSIONS

Two questions were asked at the beginning of this chapter. The first was whether there were cases in which herbivores could have a positive effect on autotrophs by speeding up nutrient recycling. The second question was more speculative and asked to what extent herbivores act like cybernetic regulators in an ecosystem through their influence on nutrient cycling.

Herbivores not only have a direct negative effect on the autotrophs they consume, but they can have indirect effects, which may be positive. A significant body of empirical research indicates that the highest production rates of plants occur when grazing occurs but is not too high. There are various possible explanations for this observed effect, one being that the

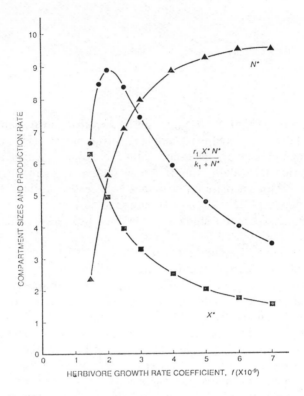

Figure 6.9 Predictions of steady-state available nutrient, N^* (μgl^{-1}), phytoplankton cell numbers, X^* ($\times 10^{-7}$ cells l^{-1}), and phytoplankton production rate from Equations (6.5) for increasing values of the herbivory coefficient, f. The parameter values are $I_n = 2.0\,\mu$gl^{-1}d^{-1}, $r_1 = 0.5$d^{-1}, $k_1 = 10.0$, $\gamma = 2.36 \times 10^{-7}\,\mu$g cell^{-1}, $d_2 = 0.04$d^{-1}, $r_n = 0.2$d^{-1} and e_1 is initially 0.10d^{-1}.

rapid recirculation of nutrients by herbivores may stimulate autotroph production.

The indirect effects of herbivore nutrient recycling can be shown even with very simple models. One model, which allows herbivore grazing and nutrient recycling levels to be adjusted to optimize conversion efficiency of radiant energy to zooplankton biomass, both predicts realistic distributions of nitrogen between ammonium, phytoplankton biomass and zooplankton biomass in the upper mixed layer of the ocean and predicts that these distributions approximately optimize phytoplankton production. Another simple model predicted that when herbivory recirculates to the available pool nutrients that would otherwise be lost from the system, per capita primary production will often be highest at relatively high levels of grazing consumption.

7 Nutrient interactions of detritus and decomposers

7.1 INTRODUCTION

In nearly all ecosystems, the majority of primary production is not consumed by herbivores, but passes directly to dead organic matter, or detritus. A smaller fraction of primary production is incorporated into herbivores and carnivores, which becomes detritus when these organisms die. The decomposition of dead organic matter is carried out by a diverse collection of organisms, ranging from microorganisms (fungi and bacteria) to relatively large animals (earthworms and millipedes). These, in turn, support an array of predators that obtain their sustenance wholly or largely from energy and nutrients stored in detritus.

Energetically, the food web based on detritus in an ecosystem may be more important than the food web based directly on autotrophic production (the grazing chain). For example, W. E. Odum (1970) worked out a food chain based on fallen mangrove leaves in southern Florida. Only about 5% of the leaves were grazed (primarily by insects) while they were alive. Fungi and bacteria colonized the dead leaves that fell in the water and a diverse group of detritivores then fed on the leaf particles and their microbial communities. These detritivores included insect larvae, nematodes, harpacticoid copepods, amphipods, crabs, snails and sheepshead minnows. As E. P. Odum (1971) pointed out, the energy that passed through the mangrove detrital food web may make a substantial contribution to the coastal fishery of southern Florida.

In addition to its importance as an energy source, detritus serves as a reservoir of nutrients. Decomposers regenerate these nutrients, so that they become available again to autotrophs. In terrestrial ecosystems nutrients released in mineral form by weathering are 'generally insignificant in relation to the nutrient demand of the vegetation. The major part of the nutrient replenishment is accomplished by the mineralization of the elements by the action of decomposer organisms. *Mineralization* is the conversion of an element from organic to inorganic form' (Swift *et al.*, 1979). Aquatic ecosystems also rely heavily on mineralization of nutrients (or remineralization if the nutrients were originally in inorganic form), particularly the phosphorus bound in phytoplankton and macrophytes. For example, Carpenter (1980) showed that the release of phosphorus from decaying submersed macrophytes may be a stimulus to pelagic primary production in lakes. In a phytoplankton culture studied by DePinto *et al.* (1986), phosphorus regeneration rates were less than $0.01 \, d^{-1}$ in cultures not inoculated with a decomposer community, but were two to five times higher for decomposer-inoculated cultures.

The purpose of this chapter is to consider the various roles detritus and the decomposer food web play in the dynamics of food webs. The size of the detritus component and the pool and nutrients held by it, controlled in part by the decomposition rate, has an important influence on food web stability. The details of the detrital/decomposer interactions also have other, more complex effects within the ecosystem.

Up to this point, the dead organic matter or detritus of an ecological system has been considered to be only a pool from which nutrients are released at a constant rate back to the pool of nutrients available for uptake by autotrophs. This simplified view overlooks the nature of detritus as a system of complex activity that is important in its own right. In this complex detrital subsystem, the return of nutrients to the available pool is not nearly as simple as described by the constant rate coefficient, d_D, used in earlier chapters. This chapter adds some complexities and explores their implications for the dynamics of food webs. This will require a more detailed description of the process of decomposition, including the explicit modelling of the decomposers and, in some cases, the heterotrophs that feed on these decomposers. Before pursuing this increase in model complexity, we will first examine, with models developed in earlier chapters, the importance of the size and turnover rate of the detrital compartment to the stability of the food web.

7.2 EFFECTS OF DETRITUS COMPARTMENT SIZE ON AUTOTROPH–HERBIVORE STABILITY

A large detrital component will usually have the effect of lowering variability of available nutrients in the ecosystem. Fluctuations in the nutrient input rates and loss rates can be buffered by nutrient releases from the large reservoir in the detrital biomass so that primary production is less variable. In view of this, it would seem that a system with a large detrital biomass component would usually be more stable than a system that has a small detrital biomass, though equivalent in other respects.

Paradoxically, model simulations show that systems with large detrital components may be associated with a greater tendency towards oscillatory behaviour than systems with small detrital components. Limit cycle oscillations are the result of local instability of a system, which was explored with respect to autotroph–herbivore interactions in Chapter 5 [Equations (5.16)]. In the analysis in Chapter 5 it was noted that local instability can result from nutrient enrichment, Rosenzweig's (1971) 'paradox of enrichment'. In systems with a large nutrient input, autotroph–herbivore interactions are more likely to produce instability and limit cycle oscillations. Thus, a reduction in nutrient input, I_n, tends to reduce oscillations in model ecosystems. When nutrient is limiting, irruptions in autotroph biomass are opposed by a decrease in the concentration of limiting nutrient. If this reactive decrease is sufficiently strong, the autotroph–herbivore oscillations may be damped.

The greater vulnerability to autotroph–herbivore limit cycles of models with large detrital components is probably due to the fact that the detritus releases nutrients into the available nutrient pool, N. This is at a rate in which N is maintained at a higher average level in this case than when the detrital component is small. The higher level of available nutrient would favour the autotroph–herbivore instability.

This proposed mechanism was tested through model simulations of hypothetical systems with small and large detrital components. The dynamics of the four variables: available nutrient, N, autotroph biomass, X, herbivore biomass, Y, and detritus, D [described by Equations (5.16)] are shown in Figure 7.1(a,b) for two values of the detrital loss rate, e_D; 1.0 and

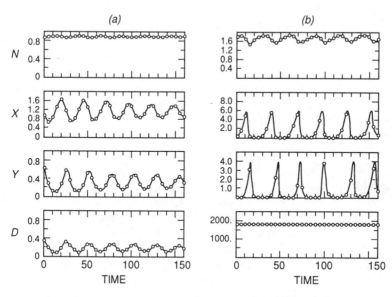

Figure 7.1   Simulations of available nutrients (*N*), autotroph biomass (*X*), herbivore biomass (*Y*) and detrital biomass (*D*) for a hypothetical system described in Equations (5.16). (*a*) Low detrital biomass, $e_D = 1.0$; (*b*) high detrital biomass, $e_D = 0.0002$. Other parameter values are: $I_n = 0.01$, $r_n = 0.001$, $k_1 = 10$, $k_2 = 10$, $d_1 = 0.06$, $d_2 = 0.5$, $e_1 = 0.0001$, $e_2 = 0.0$, $f = 5.0$ and $\eta = 1.0$, $g_1 = 0.02$, $r_1 = 2.0$, $d_D = 0.0001$, $\gamma = 0.05$. (Units are arbitrary.)

0.0002, associated with loss of nutrient from the system. Other parameter values are listed in the figure caption. Detritus was lost from the system at a higher rate in the first case than the second, so that recycling to the available nutrient pool was lower in the former case.

In the case with $e_D = 1.0$, the detrital component was small and varied as a result of the autotroph–herbivore oscillations, but these oscillations slowly damped out [Figure 7.1(a)]; whereas, when $e_D = 0.0002$, the detrital component was large and stable, but there appeared to be autotroph–herbivore limit cycle behaviour [Figure 7.1(b)]. The intense oscillations in the second case are caused by the fact that the average level of available nutrient is higher due to the larger detrital accumulation and larger fraction of nutrient recycled through the detritus.

7.3   EFFECTS OF DETRITUS COMPARTMENT SIZE ON RESILIENCE

In the preceding section, a system was examined in which autotroph–herbivore interactions are likely to lead to instability and oscillations. It was shown that a large, constant detrital reservoir of nutrients made it less likely that these oscillations would be damped by out-of-phase oscillations in available nutrients.

Let us now shift attention towards ecosystem models that are locally stable and not vulnerable to oscillations. For these systems, the resilience, or rate of return to steady-state equilibrium following a perturbation, is a key property in characterizing system dynamics. Detrital biomass plays a special role with regard to the resilience. Dudzik *et al.* (1975) performed detailed simulations of two abstract models: one a model of a shallow, mesotrophic freshwater lake and one of a grassland ecosystem. These were modelled with different levels of detail, ranging from four to eleven components (Figure 7.2). Organic litter or detritus was a component in each of these systems. One of the key generalizations Dudzik *et al.* derived from a study of their models was that: 'Perturbations in the organic litter pool can lead to more severe disturbances of the ecosystem, than the same-sized perturbations in other compartments.' In fact, in simulations of the grassland ecosystems, plant cover diminished by 50% following a 10% reduction in organic litter, and decomposers also underwent a drastic decrease. Even after a 20-year simulation, there was only a slight recovery from disturbance.

There is a logical explanation for the importance of detritus and disturbances to this component. In each of the models studied by Dudzik *et al.* (1975), organic detritus was by far the most massive compartment, holding a large fraction of the nutrient in the system. A removal of 10% of the detrital biomass in such a system, especially in a low-subsidy system where primary production relies on nutrients recycled from the detritus rather than external inputs of nutrients, can have a long-term disruptive effect.

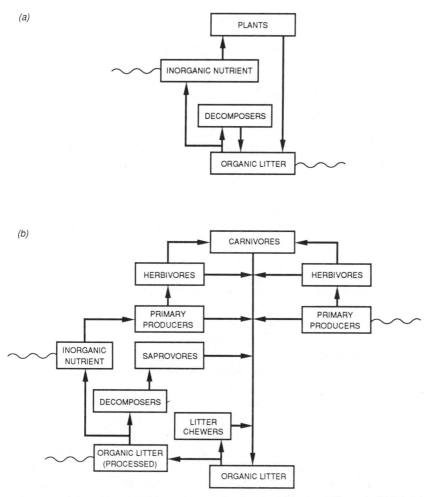

Figure 7.2   General models of nutrient cycling analysed by Dudzik *et al.* (1975). (*a*) Simple model and (*b*) complex model.

A simple model can be used to help interpret these results and to provide some general understanding of how the resilience of a system can be influenced by both the size of the detritus compartment and the nature of the disturbance. The model used is the same as in the preceding section [Equations (5.16)], except that the interaction term between the autotrophs and herbivores is changed from a Holling Type II interaction to a Holling Type III. The use of the Holling Type III function helps stabilize the system to ensure that it will return to equilibrium following a perturbation. Two widely different rates of detrital decomposition are assumed ($d_D = 0.1$ for one case and $d_D = 0.0001$ for another) so that steady-state detrital compartments of different sizes are established for the two cases. Two different types of perturbations are also applied to the system: (1) A perturbation that removes an amount (10%) only from the available nutrient N, autotroph X and herbivore Y components and (2) a perturbation that removes 10% from each of the four components. Thus, there are four different treatments. These are used in simulations to determine the effects on resilience, or inverse of the return time to equilibrium.

The return time to equilibrium, T_R [return to a fraction $e^{-1} = 0.37$ of the original displacement, averaged over the compartments as in Equation (2.8)] was calculated for each of these cases. The results are shown in Table 7.1. On the one hand, when the detritus compartment is small ($d_D = 0.1$, so the amount of nutrient it stores is of the same order as the nutrient in the other compartments), the return time to equilibrium is about the same whether only N, X and Y are perturbed or N, X, Y and D are all perturbed. On the other hand, when the detrital compartment is large ($d_D = 0.0001$, so that it stores a much greater amount of nutrient than the other components in steady state), the behaviour following a disturbance differs depending on whether the perturbation affects the detritus or not.

The above results have a simple intuitive explanation. When the detrital component was very large, a large fraction of nutrient in the system was tied up in that compartment. A decrease of 10% in the other, smaller compartments had little effect on the total nutrient in the model ecosystem. In fact, following the perturbation to the autotroph, herbivore and available nutrient components, the detrital compartment remained roughly constant. Since most of the nutrient used by the autotroph came to it via recycling from the detritus, the autotroph production continued roughly at its steady-state rate, and recovery was rapid (i.e. high resilience). When the large detrital component was perturbed by a removal of 10% of the organic matter, this caused a substantial depletion of the total nutrient in the whole system. None of the components returned to their former values until the external nutrient input had restored sufficient limiting nutrient to the system. The time required was roughly equal to the turnover time of the limiting nutrient

Table 7.1 Return times to steady-state equilibrium, T_R, following a perturbation for the system of Equations (5.16) (with Holling Type III interactions between autotrophs and herbivores) for two different detrital biomasses; corresponding to decomposition rates $d_D = 0.1$ (corresponding to small detrital biomass) and 0.0001 (large detrital biomass) and for two types of perturbations; one affecting all components and one affecting all but detrital components. The units here are arbitrary

	Decomposition rate	
	0.1	0.0001
Perturbation of N, X, Y	834.6	395.4
Perturbation of N, X, Y, D	876.4	1944.0

in the whole system. When the detrital compartment was small, a perturbation of the autotroph, herbivore and available nutrient removed a considerable fraction of the nutrient and the recovery rate of these components was limited by the relatively low input of external nutrient subsidy. The difference in results for the perturbed compartments between systems that have high and low decomposition rates can be interpreted in terms of the nutrient turnover time, $T_{res} = N_T^*/I_n$, of the system (where N_T^* is the total amount of nutrient stored in the system at steady state and I_n is the rate of nutrient input). The turnover time is greater in the system with the low decomposition rate, because a large amount of nutrient is stored in organic matter in this case.

The above analysis supports the results of Dudzik *et al.* (1975) concerning the importance of perturbations to organic litter detritus when this component contains the bulk of nutrients in the system. As can be expected by analogy, when other components of the system act as the primary storage reservoirs of nutrients in ecosystem models, the system is most sensitive to perturbations that remove fractions of those components. Dudzik *et al.* (1975), for example, showed that a model of a tropical rain forest had least resilience to perturbations that removed a fraction of autotroph biomass, since a large proportion of nutrients was stored in the woody parts of trees.

7.4 INFLUENCE OF DECOMPOSERS ON NUTRIENT RECYCLING

The simple models of earlier chapters assume not only that detritus or litter decomposes at a constant rate, but that other nutrients are released from the detritus at the same rate that the biomass decomposes and releases its main component, carbon; that is, the carbon/nutrient ratio remains constant through time. This assumption has been demonstrated to be false for many systems.

The difficulty with the assumption of constant carbon/nutrient ratio is that the activities of decomposers are themselves dependent on the ratios of nutrients in detritus and available nutrients in inorganic form. Because decomposers require nutrients, they can, when detrital nutrient concentration is low enough, act as a net sink for available nutrients in the system rather than as a net source.

Aber and Melillo (1975) found that the concentration of nitrogen in many forest litter types increases linearly as the detritus undergoes weight loss and that this trend can continue until a large fraction (at least 50 to 60%) of the original weight has been lost. This finding has been corroborated for many systems: N in floodplain and upland forests (Peterson and Rolfe, 1982; Kelly and Beauchamp, 1987); N and P in freshwater emergent macrophyte litter (Morris and Lajtha, 1986); Ca, Mg, N, P and K in plant litter in bush-fallow in sub-humid tropical Nigeria (Swift *et al.*, 1981); N in *Spartina alterniflora* litter in salt marshes (Marinucci *et al.*, 1983); N in pine litter (Berg and

Ekbohm, 1983); and P in slash pine litter in northern Florida (Gholz *et al.*, 1985). Large decaying boles in forests are often sinks for nutrient elements because of their high ratios of carbon to other nutrients. In an old undisturbed forest these boles were shown to make up 9% of the detritus (Lang and Forman, 1978). In addition to this increase in nutrient concentration in litter through time, litter has been found to decompose more rapidly when nutrients are added (e.g. Anderson, 1978).

The phenomenon of increasing nutrient concentration in litter has been explained to result from the *immobilization* of inorganic nitrogen from water solution during the microbial decomposition of the organic carbon of litter with high carbon/nutrient ratios. In this process a net gain in organic nitrogen in litter occurs (decomposers die and become part of the litter) and at the same time a net loss in organic carbon results from CO_2 evolution. The result is a decrease in the organic carbon/nutrient ratio. In this explanation, the 'organic matter' includes both the substrate for decomposition and the decomposer organisms. Addition of nutrient speeds up decomposition because the decomposers are nutrient limited.

In contrast to the decrease in the carbon/nitrogen ratio in litter when this ratio is initially high, it has been found that decomposition or organic matter with low carbon/nitrogen ratios leads to a net accumulation of inorganic or available nitrogen in soil water solution (Bartholomew, 1965). This has been attributed to mineralization of organic nitrogen during the decomposition process and its release into water solution. Addition of fertilizer sometimes also leads to the acceleration of decomposition (Harmsen and Kolenbrander, 1965). The decomposition rate of organic matter also seems to change as different substrates are added to it (e.g. Carpenter, 1981).

7.5 MECHANISTIC MODEL OF MINERALIZATION-IMMOBILIZATION PHENOMENA

To attempt to present a unified picture of these various aspects of decomposition, mathematical models have been developed, including a model by Parnas (1975). The basic idea in Parnas' model is that the decomposition rate is limited by the growth rate of decomposers. Decomposers require nutrients. Thus, their growth rate depends on nutrients available in the decaying litter and nutrient in inorganic solute form in the water. Parnas' model is purely one of the decomposition process. Decomposers are present, but the model stops short of describing their mortality and addition to the detrital biomass. Here we describe Parnas' model and its properties to see what controls mineralization and immobilization and to see how these are related to other components of the food web.

In Parnas's (1975) model detritus or organic matter, assumed to be largely plant residue, is divided into two parts: a part consisting of carbon–nitrogen, or C–N, compounds, such as protein and RNA, and a part termed C compounds, such as cellulose and starch. Detritus is then described by three variables for the standing stocks of carbon and nutrient (nitrogen in this model) in some unit area or volume.

C_1 = amount of carbon in C–N compounds;

N_1 = amount of nitrogen in C–N compounds;

C_2 = amount of carbon in C compounds.

The total detrital carbon in this area or volume is

$$C = C_1 + C_2$$

and C_1 and N_1 are assumed to occur in the C–N compounds in the fixed ratio β.

Decomposers use carbon both for the formation of structure and for providing energy. Not all carbon decomposed from detritus is assimilated into decomposer biomass. Some is released as CO_2. Define:

F = ratio of the carbon assimilated into decomposers to the total amount of carbon released from detritus by decomposers,

f_c = average fraction of carbon in the decomposer's cells,

f_n = average fraction of nitrogen in the decomposer's cells.

These definitions, plus the assumption that nitrogen is used only as structural material, imply that the ratio of carbon to nitrogen used in the formation of decomposer biomass is α/f_n, where $\alpha = f_c/F$; that is, α is the total carbon used by the decomposer organisms per unit decomposer biomass increment.

The task now is to derive equations for the changes in the carbon and nitrogen in the detritus, C_1, N_1, C_2, and the available inorganic nutrient in the area or volume, which is largely ammonium and will be represented by the variable NH_4^+ here. General equations can be written for C and N_1 that apply under all circumstances. The rate of loss of total carbon, C, from the detritus is

$$dD/dt = -\alpha GB \tag{7.1}$$

where

G = specific growth rate of decomposers, which is limited both by N and C, and

B = current biomass of decomposers.

Evidence that the decomposition rate is proportional to decomposer biomass is shown in Figure 7.3. The decomposer growth rate G is a function of total levels of carbon and nitrogen available. Parnas (1975) assumes a multiplicative Monod-type growth function (see O'Neill et al., 1989, for information on this generalization of the Monod function) for this growth

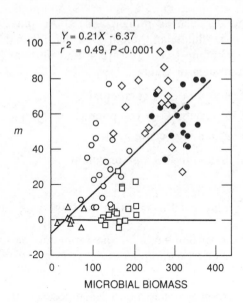

Figure 7.3 Net nitrogen mineralization, m, ($\mu g\ g^{-1}\ 20\,d^{-1}$) as a function of soil microbial biomass (mg of C per 100 g of dry soil) in soils of five sites (denoted by the different symbols) in the Serengeti grasslands. The negative values imply net nutrient immobilization. (From Ruess and McNaughton, 1987.)

rate:

$$G = G_{max}CN/[(k_c + C)(k_n + N)] \tag{7.2}$$

where

G_{max} = maximal growth rate, given high availability of carbon and nitrogen substrate

$$N = N_1 + [NH_4^+]$$

k_c, k_n = saturation constants for carbon and nitrogen.

Define further

i = immobilization rate, or net rate at which NH_4^+ in solution is taken up by microbial biomass

m = mineralization rate, or net rate of release of NH_4^+ during decomposition

One aspect of the net rate of change of nitrogen, N_1, in the detritus, is the loss due to the rate of nitrogen being taken up by the decomposer, $(f_n GB - i)$. Note that the loss from detritus of nitrogen taken up by the decomposers is offset somewhat if the decomposer obtains some of its needed nutrient directly from inorganic nitrogen in the water solution in the form of an immobilization rate, i. There is also a loss representing the amount of nitrogen mineralized per unit time, m. The equation for the rate of change of detrital nitrogen is thus

$$dN_1/dt = -(f_n GB - i) - m \qquad (7.3)$$

This is a very general equation that is specified in more detail when i and m are determined more specifically.

The rate of nutrient immobilization, i, depends on the amount of inorganic nitrogen present, NH_4^+, compared with nitrogen in the detritus, N_1. The assumption made by Parnas (1975) is that

$$i = f_n GB[NH_4^+]/([NH_4^+] + N_1) \qquad (7.4)$$

so that if $[NH_4^+] \ll N_1$ virtually all nitrogen used by decomposers comes from detritus. In the opposite limit $[NH_4^+] \gg N_1$, there is no utilization of detrital nitrogen by the decomposers, since, in that limit it can be seen that the difference between the total uptake of nitrogen and the rate of uptake of nitrogen from water, or immobilization, is very small:

$$f_n GB - f_n GB[NH_4^+]/([NH_4^+] + N_1) \ll f_n GB \qquad (7.5)$$

The remaining question is how the decomposers partition their usage of C_1 and C_2. There are two situations in which the decomposers can find themselves with respect to the resources provided by the detritus. First, they can be limited by nitrogen, when the ratio of carbon to nitrogen in detrital biomass is greater than the ratio of nitrogen used to carbon used in forming decomposer biomass, that is, when $C/N_1 > \alpha/f_n$. Second, they can be limited by carbon when the opposite inequality holds, that is, when $C/N_1 < \alpha/f_n$.

In the first case, the rates at which C_1 and C_2 are removed from detrital biomass will be set by the decomposers attempting to maximize the uptake of the limiting nutrient, nitrogen. This would include utilization of NH_4^+ from solution. Thus, the loss rate of nitrogen from detritus is

$$dN_1/dt = -(f_n GB - i) \qquad (C/N_1 > \alpha/f_n) \qquad (7.6a)$$

where the immobilization offsets some of the loss from the C–N compounds in the litter. There is no mineralization, m, because all N taken from the litter is used for the growth of the decomposers. The loss of C_1 will be proportional to this by the factor of β, defined earlier as the ratio of C_1 to N_1 in C–N compounds:

$$dC_1/dt = -\beta(f_n GB - i) \qquad (C/N_1 > \alpha/f_n) \qquad (7.6b)$$

Finally, the rate of loss of the remainder of the carbon from the detritus used to maintain the growth rate GB [Equation (7.1)] is

$$dC_2/dt = -[\alpha GB - \beta(f_n GB - i)] \qquad (C/N_1 > \alpha/f) \qquad (7.6c)$$

The ratio of removal of C_1 to C_2 can range from much greater than 1.0 to much less than 1.0. When the amount of available NH_4^+ is very large, $i \lesssim f_n GB$, and most of the carbon removed is C_2. At the other extreme, when $[NH_4^+]$ is very small, $i \ll f_n GB$, the carbon removed from C–N compounds, C_1, will play a much larger relative role, as the decomposer will be using C–N compounds preferentially over C compounds in order to obtain nitrogen. Two conclusions were derived by Parnas (1975) from these results:

> **Conclusion 1:** The C/N_1 ratio of organic material being decomposed in the absence of other nitrogen sources will increase with time if its initial C/N_1 is higher than α/f_n, and will not change with time if its C/N_1 is equal to or lower than α/f_n.
>
> **Conclusion 2:** Where the initial C/N_1 ratio in D is greater than α/f_n, the ratio can decrease with time only if other sources of N are present.

The balances of carbon and nitrogen for $C/N_1 > \alpha/f_n$ are displayed in Figures 7.4(a) and 7.4(b) respectively. In Figure 7.4(b), because nitrogen in detritus is limiting, the fluxes are determined by maximization of nitrogen uptake. Both uptake of NH_4^+ and utilization of N from C–N compounds in

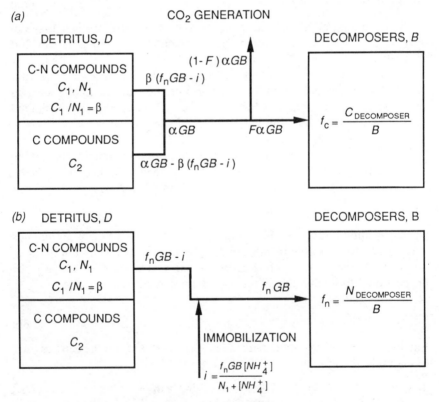

Figure 7.4 Fluxes from detritus to decomposers of (a) carbon and (b) nitrogen in Parnas' (1976) model, when nitrogen is limiting to decomposers. There is an immobilization rate, i, of NH_4^+ from the solute. $C_{\text{decomposer}}$ and $N_{\text{decomposer}}$ are the amounts of carbon and nitrogen in the decomposers.

the detritus will be used to obtain a total flux of nitrogen, $f_n GB$, to the decomposers. This fixes the flux of carbon from the C–N compounds as $\beta(f_n GB - i)$ (Figure 7.4a). The carbon flux from the C compounds in the detritus makes up the difference.

In the second case, when carbon is limiting ($C/N_1 < \alpha/f_n$), the decomposition rate, αGB, is assumed divided between the two fractions in the detritus;

$$dC_1/dt = -\alpha GBC_1/C \quad (C/N_1 < \alpha/f_n) \tag{7.7a}$$

$$dC_2/dt = -\alpha GBC_2/C \quad (C/N_1 < \alpha/f_n) \tag{7.7b}$$

Now N_1 is removed from the detrital biomass in proportion to C_1, or

$$dN_1/dt = -\alpha GBN_1/C \quad (C/N_1 < \alpha/f_n) \tag{7.7c}$$

The decomposer is using the C–N compounds foremost as a carbon source and not all of the nitrogen removed in the process is used. The excess nitrogen may be mineralized. Figure 7.5(a,b) shows the fluxes of carbon and nitrogen between the detritus and decomposers in this case.

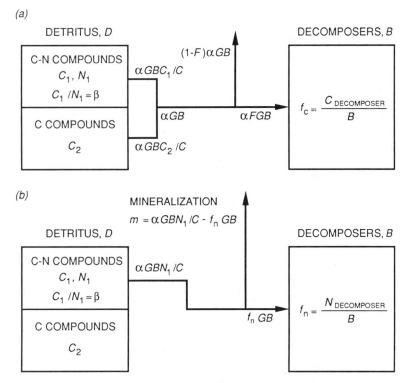

Figure 7.5 Fluxes of (*a*) carbon and (*b*) nitrogen in Parnas' (1975) model, when carbon is limiting to decomposers. There is a net mineralization rate, *m*, of nitrogen.

It is now possible to compute the instantaneous rate of mineralization m as a function of the key variables of the system. By using the general and always valid Equation (7.3), an expression can be derived for m;

$$m = -(f_n GB - i) - dN_1/dt \tag{7.8}$$

When carbon is limiting $(C/N_1 < \alpha/f_n)$, dN_1/dt is given by Equation (7.7c) and the immobilization rate $i = 0$. Then, the mineralization rate is

$$m = GB[(\alpha/C)N_1 - f_n] \tag{7.9}$$

Parnas (1975) drew the following conclusions:

> **Conclusion 3:** Mineralization occurs when the C/N_1 ratio for D being decomposed is smaller than α/f_n.
> **Conclusion 4:** The rate of mineralization is increased as C/N_1 is decreased.
> **Conclusion 5:** The higher the amount of NH_4^+ in the area, the higher will be the rate of mineralization.

This rate of mineralization can now be calculated numerically for various values of N_1, C_1, C_2 and $[NH_4^+]$, using Equation (7.9) plus Equation (7.2). Parnas (1975) determined numerical values for parameters of the model as part of a larger model for the US International Biological Programme's Desert Biome (Table 7.2). For these parameters, m is calculated as a function

Table 7.2 Parameter values for a model of decomposition of organic matter by decomposers (from Parnas, 1975)

Parameter	Meaning	Value
G_{max}	Maximal growth rate of a mixture of decomposers	$0.01 \, d^{-1}$
k_c	Constant in Michaelis–Menten term that is the concentration of carbon for half the maximum growth rate of the decomposer	$20 \, g \, m^{-2}$
k_n	Constant in Michaelis–Menten term that is the concentration of nitrogen for half the maximum growth rate of the decomposer	$0.4 \, g \, m^{-2}$
F	Efficiency of carbon assimilation	0.4
f_c	Fraction of carbon in microbial biomass	0.5
f_n	Fraction of nitrogen in microbial biomass	0.05
β	Molecular ratio between carbon and nitrogen in protein	4.0

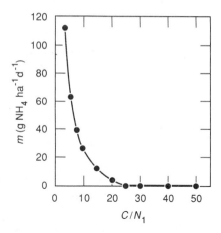

Figure 7.6   Mineralization rate, m, as a function of the C/N_1 ratio in detritus based on parameter values from Table 7.2. (From Parnas, 1975.)

of C/N_1 (Figure 7.6). Below a ratio of $C/N_1 = \alpha/f_n = 25$, mineralization occurs and grows exponentially as C/N_1 decreases.

Parnas (1975) model agrees well with experimental results. For example, in experiments in which organic matter decays and the only nitrogen is that in the C–N compounds in the organic matter, the C/N_1 ratio tends to increase with time (Brown and Dickey, 1970). This finding is predicted by the model, because the C–N compounds are used at a faster rate (proportional to β) than the C compounds (rate proportional to $\alpha - \beta f_n$). When there is additional available nitrogen, then the C/N_1 ratio can decrease due to nitrogen immobilization. The model also supports the experimental finding that when the decaying organic matter is very rich in nitrogen, mineralization will occur. Addition of NH_4^+ will increase the rate of mineralization (Broadbent and Nakashima, 1971).

The equations derived above for the decomposition of detritus do not represent a complete description of the system, since there are no terms for the production of new detritus, no terms for the loss of biomass of the decomposers back to detritus, and no equations for the available nutrient pool or the trophic levels in the ecosystem. In order to provide a description of the effects of decomposition on food web properties, the decomposition model must be integrated into a more complete model of the system. I will not attempt to complete the description of a basic ecosystem here, but will note that such a description must include an equation for available nutrient.

$$d[NH_4^+]/dt = I_n - r_n[NH_4^+] + m - i - F_{NX} \qquad (7.10)$$

where an external input, I_n, and a loss, $r_n[NH_4^+]$, have been included and the last term, F_{NX}, represents nutrient uptake by the autotroph, an equation for autotroph biomass, X, an equation for decomposer biomass, B, and equations for C_1, C_2 and N_1. The equations for X and B must reflect the fact that when there is net immobilization of available nutrients (i.e. when $i > m$) the decomposers and autotrophs compete for nutrients. The autotroph uptake term, F_{NX}, and the immobilization rate, i, must take into account competition between the decomposers and autotrophs for nutrient uptake.

7.6 EFFECTS OF HIGHER TROPHIC LEVELS IN THE DETRITAL FOOD CHAIN

The early concepts of decomposition and recycling of nutrients pictured a system in which bacteria directly released mineralized nutrients into the available pool. However, as discussed above, bacteria may also immobilize nutrients under some circumstances. It has frequently been noted that herbivorous grazing on the bacteria either speeds up the mineralization of nutrients or may even be necessary for net mineralization to occur.

Johannes (1965, 1968) showed in aquatic and marine systems that regeneration of inorganic phosphate from organic detritus proceeded faster when both bacteria and ciliates (which feed on bacteria) are present than when only bacteria are present. Bacteria tended primarily to immobilize nutrients. Barsdate et al. (1974) used aquatic microcosms and showed that phosphorus cycled much more rapidly in systems with bacteria and protozoan grazers than with bacteria alone. They hypothesized that the turnover rate of phosphorus in grazed bacteria was higher than that of ungrazed bacteria.

Buechler and Dillon (1974) studied phosphorus turnover by Paramecia, which graze on bacteria. They found that bacteria are efficient in removing P from media. They tend not to release it. Paramecia feed voraciously on bacteria and excrete most of it. This may account for the release of large amounts of P back into the available pool. Gallepp (1979) noted, in sediment–water microcosms, that chironomids appear to increase the concentration of P in the water above sediments. Chironomids are filter feeders and they release P in faeces.

Cole et al. (1978) made similar studies on terrestrial microcosms, simulating rhizospheres with combinations of bacteria, amoeba and nematode populations. Bacteria alone immobilized much of the P from the detritus. Because of the high consumption rate and relatively low assimilation rate of the grazers, amoebae mineralized much of the bacterial P, returning it to an available inorganic pool. At a higher trophic level, nematodes decreased inorganic P, possibly because it limited amoebae.

Anderson et al. (1983) carried out a series of microcosm experiments with leaf litter and soil organic matter from deciduous woodlands. Different levels of soil grazing on fungal biomass by macrofauna (e.g. millipedes) were added for a period of 12 weeks and levels of NH_4^+N and NO_3^-N were measured in the soil. Addition of grazers resulted in reduction in nutrient immobilization in litter and soil organic matter.

Douce and Webb (1978) modelled the indirect effects of soil invertebrates on litter decomposition. One effect of soil invertebrates is to graze bacteria (microflora) down to the level where they are growing exponentially. This shortens turnover times. Analogously, Schaeffer and Whitford (1981) proposed that higher trophic levels release nutrients stored by termites. Ants, lizards and birds feed on termites and return nutrients to surface soil.

Ingham et al. (1985) proposed a conceptual model for the flows of nutrients (nitrogen and phosphorus) in the decomposer web (Figure 7.7). Note that in this system not only bacterial, but also fungal, decomposition is represented, along with grazers (nematodes) feeding on each. In a set of experimental studies, Ingham et al. started with sterile soil and added various components of the decomposer web to see how various combinations affected nutrient mineralization and, hence, plant growth rate. In particular, they tested the hypothesis that nematodes feeding on fungi or bacteria increased the amount of inorganic nutrients available for plant uptake. [This concept is very similar to the effect of herbivores on nutrient levels examined in Chapter 5 (Figure 5.5), where herbivory controlled the autotroph component, which allowed available nutrients to build up.]

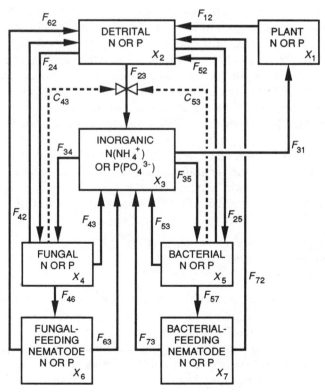

Figure 7.7 Conceptual model of the nutrient flows in a plant plus decomposer system. (From Ingham *et al.*, 1985.)

Ingham *et al.* (1985) found in one set of studies that autotroph production (shoot growth) was higher with bacterial-feeding nematodes than with bacteria alone. Interestingly, bacterial numbers were also higher in the treatment with nematodes than without. The authors proposed a few hypotheses for the apparent stimulatory effect of nematodes on bacteria, including non-lethal passage of many bacteria through nematode guts, where they are exposed to high nutrient levels. This possibility is analogous to the stimulations of autotroph growth by herbivores discussed in Chapter 5.

The interactions of larger grazers with decomposers have even more mutualistic overtones. Ruess and McNaughton (1987) examined nutrient dynamics in different areas of the Serengeti grasslands that are exposed to different levels of ungulate grazing: tallgrass (low grazing), midgrass (moderate grazing) and shortgrass (high grazing) sites. They found that microbial biomass was positively related to the grazing rate. As Figure 7.3 shows, the net rate of mineralization, m, is positively related to microbial biomass, and, as well, is inversely related to the C/N_1 ratio:

$$m = 0.19B - 3.19(C/N_1) + 40.48 \qquad (r^2 = 0.55, \ P < 0.0001) \tag{7.11}$$

Both microbial biomass and mineralization were higher on highly grazed sites. Hence grazing appears to increase soil microbial activity and mineralization in the same sense as it does net primary production and plant net nutrient flux (Ruess *et al.* 1983; Ruess, 1984; McNaughton and Chapin, 1985). The dung from grazers has readily available nutrients and carbon. In some sense the interaction can be thought of as a mutualistic one, since the vertebrate rumens provide an ideal environment for microbes.

Another example of symbiosis is that between fungus-growing ants and some species of saprophytic fungi. The ants cut live leaf parts, prepare them in a variety of ways and place them in fungus gardens. The fungi degrade the plant structural carbohydrates and provide the ants with available nutrients and energy (Swift *et al.*, 1979).

Visser (1986) reviewing the influences of soil invertebrates on microbes, mentioned three main effects:

1. Comminution, or mixing and channelling of litter and soil.
2. Grazing on the microflora.
3. Dispersal of microbial propagules.

Of these, comminution has traditionally been felt to be the most important in speeding nutrient recycling, through 'exposing a greater surface area to microbial attack' (Satchell, 1974). This may not be the case for all soil invertebrates, however. Hassall *et al.* (1987) were unable to substantiate that comminution of leaf litter enhanced microbial metabolism. They suggested instead that the foraging of isopods on leaf litter on the surface at night and transporting the material as faeces to moist resting sites may accelerate soil microbe use of the material.

7.7   SUMMARY AND CONCLUSIONS

The dead organic matter or detritus can play an important role in the nutrient dynamics of an ecosystem and, therefore, has implications for system stability. In particular, a large detrital compartment buffers the living components of the system against large fluctuations in nutrient availability. This does not necessarily increase all types of stability, however. For example, we found in Chapter 5 that limitations in nutrient availability generated by herbivore–autotroph cycles tend to counteract those cycles, often stabilizing the system. In the present chapter it was shown that when the system has a large, relatively constant detrital component, these nutrient fluctuations do not occur and herbivore–autotroph oscillations will not be damped by out-of-phase changes in nutrient availability.

In contrast to this possible effect of not hindering autotroph–herbivore oscillations, a detrital component that is large relative to the living components of the system can have a positive effect on resilience of the system to perturbations affecting the living parts of the system. The steady local input of the nutrient from the decomposing detrital compartment enables autotrophs and higher trophic levels to recover more quickly than they would if only external sources of limiting nutrients had to be relied on. It was also shown, however, that a perturbation to the detrital compartment in such a case, as might occur in a stream during a heavy scouring event, could have a drastic effect on the system, as resilience would be very low for such a disturbance.

This chapter has also touched on some of the actual complexity involved in decomposition and nutrient release, which in previous systems was assumed to occur at a rate linearly proportional to detrital compartment size. In fact, detritus is a mixture of compounds with different ratios of nutrients such as nitrogen and phosphorus to carbon. The decomposers that break down detrital matter need nutrients also, and this affects how rapidly they act and whether they immobilize the nutrients released from the detritus in their own biomass or cause a net mineralization of nutrients, which are then available for the autotrophs of the system. The rate of nutrient mineralization is not controlled by the decomposers, bacteria and fungi, alone, but is affected in a variety of ways by soil invertebrates and herbivorous vertebrates.

8 Nutrient limitation and food webs

8.1 INTRODUCTION

Herbivores constituted the highest trophic level in the models of the preceding chapters. Most food webs also contain carnivores that feed on the herbivores, and often second-order carnivores that feed on the first-order carnivores. In principle, chains with any number of trophic levels may exist within ecosystems. Empirical evidence, however, indicates that in nature the number of trophic levels within ecosystems seldom exceeds five or six. In an analysis of 56 ecosystems whose food webs have been characterized in detail, Pimm (1982) showed that the great majority (47) had three or four levels, while only five and two, respectively, had five and six trophic level chains.

The purpose of this chapter is to study the effects of nutrient cycling in food chains in general, which in principle will mean food chains of any length. In practice that may mean food chains that include first-order and maybe second-order carnivores. A few words should be repeated regarding the difference between food webs and food chains noted in Chapter 1. In a food chain each trophic level is well defined; the species on that level all feed only on species in the trophic level just below it and are fed on at most by species in the level just above it. In most ecosystems, the trophic relationships among the species are seldom so clear. A second-order carnivore, for example, might feed on herbivores as well as on first-order carnivores (see Figure 1.5 for a typical food web diagram). Thus, theory must be able to deal with food webs as well as food chains. Food chains, however, are a better starting place for understanding the complex interactions that can occur between different levels and for deriving the implications that nutrient limitation and recycling have for these interactions and vice versa. Many of the examples in this chapter are food chains but some of the results can be generalized beyond chains to food webs.

One of the most difficult tasks of ecological theory is to explain the observed structure of food webs. If food chains of any length can be imagined, why do most of them have no more than three or four levels? There are a number of different hypotheses for this observation.

The first hypothesis is based simply on energy limitation. Energy is transferred up the food chain through consumption of herbivores on autotrophs, carnivores on herbivores, and so forth. The great majority of energy is lost between transfers, however (e.g. Hutchinson, 1959). All plants and animals respire and animals produce urine and faeces. Much animal and plant biomass simply becomes detritus and is decomposed by bacteria. In addition, during the process of consumption and assimilation, the Second Law of Thermodynamics ensures that the energy transfer is not perfectly efficient. The fraction of energy that does pass up the food chain from one trophic level to the next, or the 'ecological efficiency', may typically constitute less than 10% of the production of the lower level. This means that production will decrease with trophic level, n, as $(0.1)^n$. For instance, in very rough terms, more than a square kilometre of area would be required for the same rate of production of first-order carnivores as could be attained by autotrophs on a hectare. Higher trophic level animals therefore have to obtain food over large areas, which itself incurs high energy costs of travelling. Although it is not uncommon for some predators to feed over areas of hundreds or even thousands of square kilometres, energy costs may become prohibitive for continuous movements over areas much beyond that range. The arguments for energy limitation of food chain length have been developed in detail through mathematical models by Yodzis (1981, 1984).

One possible problem with the arguments for limits on food chain length based on energetics was pointed out by Pimm and Lawton (1977) (see also Pimm, 1982). They noted that there were no significant differences in the numbers of trophic levels in ecosystems having very different primary production levels. For example, three trophic levels are present in tundra regions averaging $440 \, \text{mg} \, \text{C} \, \text{m}^{-2} \text{d}^{-1}$ (Wielgolaski, 1975), while no more than four trophic levels were noted in a fish pond with a productivity of $4000 \, \text{mg} \, \text{C} \, \text{m}^{-2} \text{d}^{-1}$ (Dykyjova and Kvet, 1978). Pimm and Lawton proposed the alternative hypothesis that dynamic properties of food chains make long chains improbable. In particular, the authors simulated models of food chains of various lengths and showed that longer ones had lower resilience than shorter chains. In other words, following perturbations, a long food chain would take a longer time to recover than a short one. This would put the highest level species in a food chain at high risk, because its food source would be unstable.

The results for the resilience of autotroph–herbivore systems discussed in Chapter 5, where the herbivore limited the autotroph standing stock, may have some relevance to the resilience argument for shorter food chains. In Chapter 5 it was shown that the nutrient-limited autotroph–herbivore chain could actually be more resilient than the situation in which the herbivore was absent. This could occur because the effect of the herbivore was to decrease the turnover time of nutrient in the system. This finding does not seem to have general implications for Pimm and Lawton's hypothesis, however. Beyond two trophic levels, the resilience of food chains becomes much more complex and in general resilience does seem to increase with food chain length, according to my own simulations of long food chains with non-linear trophic interactions (see also the results later in this chapter for food chains with linear trophic interactions). Even the results of higher resilience for the autotroph–herbivore system shown in Chapter 5 depend on there being top-down regulation of the autotroph by an herbivore with a faster turnover rate.

Some other explanations of observed food web length have been proposed, but will not be reviewed here (see Pimm, 1982). The subject remains a controversial one and not all ecologists are convinced that food chains are, in fact, typically short. In studying detrital-based food webs in the soil of semi-arid grasslands, Ingham *et al.* (1986) found chains of up to seven links that appeared to be quite stable. These chains existed within highly complex webs.

The problem of explaining observed food chain lengths is only one of many problems that food web ecologists puzzle over. It is an interesting one from the viewpoint of nutrient cycling because nutrient limitation and cycling may play a role along with energy in food web resilience and structural aspects such as food chain length. This will be examined below, along with some other effects of nutrients on food webs.

8.2 NUTRIENT INFLUENCES ON DIRECTION OF FOOD CHAIN CONTROL

In Chapter 5 simple food chains with autotrophs and herbivores were discussed. In particular, models were used in which autotroph biomass was held to a constant level by the herbivores, while both nutrient concentration and herbivore biomass increased with increasing nutrient input, I_n, to the system (Figure 5.5), although the herbivore biomass approached an asymptote rather than increasing indefinitely. This 'top-down' control on the autotroph in an autotroph–herbivore system is actually a special case of a more general phenomenon that has been termed 'cascading interactions' (Carpenter *et al.*, 1985). The idea was also discussed much earlier by Hairston *et al.* (1960) for autotroph–herbivore–carnivore systems. This is an appropriate place to contrast this phenomenon of cascading, or top-down, control with resource, or bottom-up, control in food webs, before asking how nutrients affect this picture.

Bottom-up control in food chains emphasizes the importance of nutrient and light availability for primary producers and the subsequent energy and nutrient flow through a series of trophic levels in affecting the sizes of trophic levels in the food web and their dynamics. This paradigm has a long history of support among ecologists (e.g. Elton, 1927; Lindeman, 1942) and continues to be emphasized as a major factor controlling the number of trophic levels and the standing crop of organisms in ecosystems (e.g. Yodzis, 1984; Wetzel, 1983; Harris, 1986).

To help discuss the differences, let us consider Figure 8.1, which portrays nutrient movement in a nutrient-limited food chain. In this model, N_x, N_y,

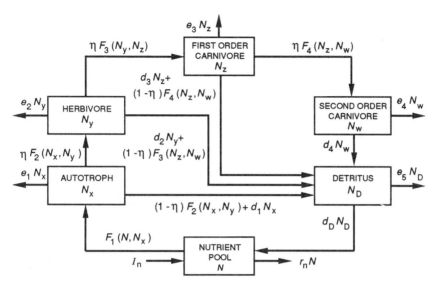

Figure 8.1 A simple four-trophic-level food chain. The arrows represent flows of nutrients through the system. Compartments for the available nutrient pool and detritus are also included.

N_z and N_w represent the equivalent nutrient standing stocks associated with the autotroph, herbivore, first-order carnivore and second-order carnivore biomasses respectively. Thus the variables differ from earlier chapters, where the variables X and Y were biomasses. The present notation is used for simplification, to avoid having to keep track of different ratios of nutrients γ_x, γ_y, γ_z and γ_w, in each trophic level. There is no effect on the model dynamics of this change in notation. The function $F_1(N, N_x)$ represents the nutrient-limited rate of primary production, while $F_2(N_x, N_y)$, $F_3(N_y, N_z)$ and $F_4(N_z, N_w)$ are production rates of the herbivore, first-order carnivore and second-order carnivore, based on their trophic interactions with the next lower trophic level. The d_i values are rate coefficients of flux to detritus due to biomass loss and mortality not associated with consumption. The e_i values are rate coefficients of losses out of the system from the various trophic levels. The constant η, represents the fraction of consumed biomass that is assimilated, the rest passing directly to detritus.

The forms of the production rates, $F_1(N, N_x)$, $F_2(N_x, N_y)$, $F_3(N_y, N_z)$ and $F_4(N_z, N_w)$ determine the type of control. Consider the linear case when the $F_1(N, N_x) = f_1 N$, $F_2(N_x, N_y) = f_2 N_x$, $F_3(N_y, N_z) = f_3 N_y$ and $F_4(N_z, N_w) = f_4 N_z$, where the f_i values are constants. This case is called 'resource controlled' or 'donor controlled' because the flux from each component of the food web depends only on the level of that component, not on the level of the component that consumes it. Assume also, to simplify calculations, that $\eta = 1$. This means that there is total assimilation of consumed nutrients from one trophic level by the higher level. The equations that result for these assumptions are

$$dN/dt = I_n - (r_1 + r_n)N + d_D N_d \tag{8.1a}$$

$$dN_x/dt = r_1 N - (d_1 + e_1 + f_2)N_x \tag{8.1b}$$

$$dN_y/dt = f_2 N_x - (d_2 + e_2 + f_3)N_y \tag{8.1c}$$

$$dN_z/dt = f_3 N_y - (d_3 + e_3 + f_4)N_z \tag{8.1d}$$

$$dN_w/dt = f_4 N_z - (d_4 + e_4)N_w \tag{8.1e}$$

$$dN_d/dt = d_1 N_x + d_2 N_y + d_3 N_z + d_4 N_w - (d_D + e_D)N_d \tag{8.1f}$$

When this model is solved for the steady-state equilibrium values, the results are

$$N^* = I_n/(r_n + r_1 - K) \tag{8.2a}$$

$$N_x^* = r_1 N^*/(f_2 + d_1 + e_1) \tag{8.2b}$$

$$N_y^* = f_2 N_x^*/(f_3 + d_2 + e_2) \tag{8.2c}$$

$$N_z^* = f_3 N_y^*/(f_4 + d_3 + e_3) \tag{8.2d}$$

$$N_w^* = f_4 N_z^*/(d_4 + e_4) \tag{8.2e}$$

$$N_d^* = (d_1 N_x^* + d_2 N_y^* + d_3 N_z^* + d_4 N_w^*)/(d_D + e_D) \tag{8.2f}$$

where

$$K = [d_D/(d_D + e_D)][r_1/(f_2 + d_1 + e_1)](d_1 + [f_2/(f_3 + d_2 + e_2)]$$
$$\{[d_2 + f_3/(f_4 + d_3 + e_3)][d_3 + d_4 f_4/(d_4 + e_4)]\}) \tag{8.2g}$$

The linear donor-controlled trophic interaction model then predicts that each of the trophic levels is linearly dependent on the input of the limiting nutrient, I_n. This shows the effect of bottom-up control on every one of the trophic levels. There is no reason that the system must be linear. If the loss rates are proportional to the squares of the compartment sizes, $-(d_1 + e_1 + f_2)N_x^2$ for example, there will still be bottom-up control for all of the higher trophic levels. However, the strength of the bottom-up effect will diminish up the food chain. In this particular case, the nutrient standing stock of the second-order carnivore will scale roughly as $(I_n)^{1/16}$. Density dependence of this sort tends to buffer the higher trophic levels against fluctuations occurring in the lower levels.

This is an interesting result in view of nutrient enrichment experiments performed by Hurd and Wolf (1974). The authors subjected two old fields to pulses of N, P, K fertilizers early in the growing season of 1970 and then monitored arthropod consumers over two years. They concluded that 'the magnitude of deflections caused by the enrichment was decreased up the trophic chain from the herbivore to the carnivore levels'.

In this model the higher trophic levels affect the lower trophic levels through recycling. For example, an increase in recycling efficiency by any trophic level such that losses from the system are decreased (i.e. any decreases in e_i with respect to d_i) will cause K to increase, which [from Equations (8.2)] causes all of the steady-state values to increase. This recycling effect of higher on lower trophic levels does not, however, constitute top-down control in the sense that the term is generally used. Top-down influence and sometimes control is exerted when the trophic interaction functions depend on the size of the consumer compartment as well as that of the resource compartment. This will be discussed next.

The model represented by Equations (8.1a–f), although convenient to analyse because the trophic interactions are linearly dependent on resource levels, is generally not in accord with reality. The rationale for more complex trophic interaction or production functions, such as the Monod growth function, $r_1 N N_x/(k_1 + N)$, for autotrophs and the Holling Type II and III functions for herbivore and carnivore consumption, were given in Chapters 3 and 5. Assumption of these production rates changes the nature of the food chain from a resource controlled system to one in which there can also be some degree of control from the top down. The effect of a Holling Type II function for second-order carnivore feeding on the first-order carnivores can be seen by substituting $f_4 N_z N_w/(k_4 + N_z)$ in place of $f_4 N_z$ in Equations (8.1d,e). At steady state, when $dN_w/dt = 0$

$$N_z^* = k_4(d_4 + e_4)/(f_4 - d_4 - e_4)$$

Thus, N_z^* is held fixed by the second-order carnivore.

The potential importance of consumer or top-down control was recognized three decades ago. Hairston et al. (1960) hypothesized that many terrestrial ecosystems exemplify top-down control. The authors specifically had in mind the control of first-order carnivores on herbivores. Because the biomass of herbivores is held to a relatively low level, according to the hypothesis, the autotrophic level is able to increase to large biomass levels if nutrient availability is enhanced. This hypothesis, if correct, would explain why plant biomass is so high in many of the earth's terrestrial systems when sunlight, water and nutrients are available in abundance. The hypothesis of Hairston et al. has been generalized to food chains that involve several trophic levels (Fretwell, 1977; Oksanen et al., 1981). According to the more general version, whatever trophic level is the top level will normally control the one beneath it (that is, regulate its standing stock of biomass), as we showed in Chapter 5 [Equation (5.5a) or (5.12b)] for control of autotrophs by herbivores. Implicit in this hypothesis is the assumption that the top trophic level occurs more than just sporadically in the system. It must be present to a significant enough degree to control the level below it. Consider Figure 8.1, for example. If the second-order carnivore, N_w, controls the first-order carnivore, N_z, to some level N_z^*, then the herbivores, N_y, are not limited by carnivores, N_z and can increase to the greatest extent of their limitation by resources (energy and nutrients) from the autotrophs. The autotrophs are controlled from above by N_y, like the first-order carnivores are by N_w. The herbivores and second-order carnivores are controlled from below. Thus the alternation of regulation (from above or below) could, in theory, continue all the way down to the abiotic resources at the bottom of the chain. This pattern of regulation contrasts with the bottom-up regulation in the system of Equations (8.1).

A number of experiments have been carried out (often in microcosms) to test the hypothesis that the direction of control of autotrophs in the system depends on the number of trophic levels above the autotroph level. Elliott *et al.* (1983) established four different trophic chains: an autotroph alone [algae (A)], autotroph–herbivore [algae–zooplankton (AZ)], autotroph–herbivore–carnivore [algae–zooplankton–fish (AZF)] and a variation on the last in which the fish were regulated. It was found that algal standing stocks were high when food chains were odd (A, AZF). In both of these cases, the primary production was controlled from the bottom-up by nutrients.

Levitan *et al.* (1985) similarly showed that in an even-length chain (algae and *Daphnia*) the *Daphnia* was able to suppress the growth of the algae, even when the algae were stimulated by high nutrient levels. When carnivores in the form of trout were introduced in large numbers, they held *Daphnia* numbers down, so that in this case nutrients led to an increase in algal biomass. Similar results were found by Hessen and Nilssen (1986). In contrast, however, Walters *et al.* (1987), while observing herbivore control of primary producer biomass in oligotrophic systems, did not observe top-down control of zooplankton by fish or invertebrate predators when higher nutrient conditions were created through fertilization. Thus, the hypothesis of top-down control is a fairly good generalization, but one that may depend on circumstances and not be universal.

Altogether, there has been a great deal of support of control from the bottom up (summarized by White, 1978) as well as for control from the top down (exemplified by Mech *et al.*, 1971 and Simenstad *et al.*, 1978), but data have seldom been complete enough for a thorough analysis of a system. However, recently, sufficient data have been synthesized for a detailed examination of the direction of control in freshwater pelagic ecosystems (McQueen *et al.*, 1986). These authors compiled empirical evidence from the literature, where it exists in the form of regression relationships of the four following types: total phosphorus versus chlorophyll *a*; zooplankton biomass versus chlorophyll *a* or phytoplankton biomass; planktivore biomass versus zooplankton biomass; piscivore biomass versus planktivore biomass.

The regressions came from two basic sources. The first source is regression relations of these variables across a variety of lakes with a range of different trophic statuses. McQueen *et al.* (1986) reviewed more than a dozen such studies and showed that the mean regression lines indicate positive slopes in all cases except that of the piscivore versus planktivore indices; that is, an increase in the resource level (phosphorus) is associated with an increase in all trophic levels but the highest (Figure 8.2*b*). The second source

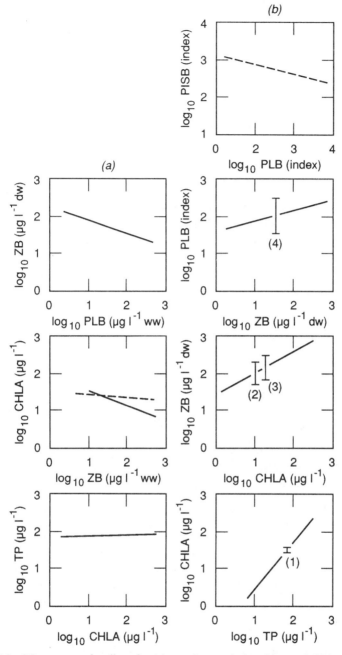

Figure 8.2 Mean regression lines for (*a*) top-down relationships and (*b*) bottom-up relationships. Numbered 95% confidence intervals are based on data from Janus and Vollenweider (1981), Mills and Schiavone (1982), and Pace (1984). The symbols are: TP, total phosphorus; CHLA, chlorophyll *a*; ZB, zooplankton biomass; dw, dry weight; ww, wet weight. Chlorophyll *a* is a good index of living autotrophic biomass. PLB, planktivore biomass (index); PISB, piscivore biomass (index). (From McQueen *et al.*, 1986.)

of data is from lakes in which 'biomanipulation' (Shapiro et al., 1982; Shapiro and Wright, 1984) has been used, that is, in which additions or removals of a top consumer are deliberately made to attempt to regulate the lower levels. The mean trends of these data are shown in Figure 8.2(a). A top-down effect of planktivore biomass on zooplankton is evident in the top frame in Figure 8.2(a). The top-down effect of zooplankton on phytoplankton is weak, except in the case when the zooplankton are cladocerans [solid line in Figure 8.2(a), middle frame]. There is little effect of phytoplankton on total phosphorus (bottom frame). The findings of McQueen et al. (1986) support, in part, the arguments of Hairston et al. (1960). They indicate that the herbivores (zooplankton) are negatively affected by carnivores (planktivores) and also, in agreement with Hairston et al., (1960), that autotrophs (phytoplankton) are only weakly affected by herbivores, except in oligotrophic lakes or where large Daphnia are present, in which case the downward affect can be strong.

To what extent can these results be explained by simple food web mechanisms in models? Cascading interactions can be presented in a somewhat general way through consideration of the food chain in Figure 8.1. I will ignore the top trophic level, N_w, for now so that the first-order carnivore, or planktivore, is the highest level. For present purposes, let the primary production function be the Monod growth function

$$F_1(N, N_x) = r_1 N N_x/(k_1 + N)$$

Think of the trophic levels N_x, N_y and N_z as potential invaders of a system that initially contains no biomass, only an available nutrient pool at a steady-state level of $N^* = I_n/r_n$. Suppose the rate of external nutrient input, I_n, is controlled externally and that it is gradually increased from zero. For low enough values of I_n, specifically, for

$$I_n < r_n k_1(d_1 + e_1)/(r_1 - d_1 - e_1) \tag{8.3}$$

the autotroph level cannot invade, because the steady-state level of N^* is too low. As I_n increases, N^* increases until the inequality is reversed (Figure 8.3). As we have seen in earlier chapters, when the autotroph invades and achieves a steady state, N_x^*, an interesting thing happens. The autotroph exerts control on the nutrient level and holds it at a fixed value

$$N^* = k_1(d_1 + e_1)/(r_1 - d_1 - e_1)$$

Further increases in I_n cause only the autotroph level to increase, which occurs until the herbivore level can invade, at which point N_x^* is held by the herbivore to a level

$$N_x^* = b_2(d_2 + e_2)/(\eta f_2 - d_2 - e_2)$$

assuming that the feeding relation between the herbivore and autotroph is expressed by a functional form that is reasonable for this interaction, the

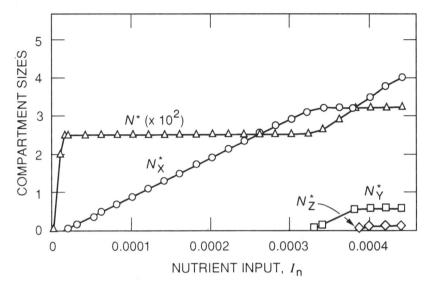

Figure 8.3 Changes in levels of available nutrient level, N^*, autotroph biomass, N_x^*, herbivore biomass, N_y^* and carnivore biomass, N_z^*, as functions of increasing nutrient input, I_n. Other parameter values are $r_n = 0.005$, $r_1 = 0.3$, $k_1 = 0.005$, $k_2 = 60$, $k_3 = 2$, $f_1 = 2$, $f_2 = 10$, $\eta = 0.5$, $d_1 = 0.1$, $d_2 = 0.01$, $d_3 = 0.5$.

Holling Type II function

$$F_2(N_x, N_y) = f_1 N_x N_y/(b_2 + N_x)$$

To see where this value N_x^* comes from, note that the right-hand side of the equation for the herbivore implied by Figure 8.1 is $[f_2 N_y N_x/(b_2 + N_x)] - (d_2 + e_2)N_y$. When this is set to zero (for steady state) and solved for N_x^*, the above result is derived.

The same pattern that occurred with the autotroph and herbivore repeats itself with each new invasion by a trophic level. When I_n increases enough, the level of N_y^* is large enough to support the invasion of the first-order carnivore level. The herbivore is then controlled by the first-order carnivore, while the autotroph is now not controlled from above and can control the available nutrient level in the system. Thus the highest trophic level and the input I_n play complementary roles. The nutrient input I_n must be large enough to allow a given trophic level to invade. Each invasion by a new trophic level regulates the sizes of the trophic level immediately below it.

How far up and down the chain do the bottom-up and top-down effects extend? There is eventual diminution of the effect in both directions. Note in Figure 8.3 that, as I_n is increased, only N^* or, alternatively, N_x^*, can continue to increase indefinitely. Higher trophic levels eventually approach asymptotes with increases in I_n. Thus, there is a limit beyond which increasing I_n does not have much influence on higher trophic levels. This effect results because of inherent saturation effects in Monod and Holling growth functions.

The possible top-down effects in the above model can be compared with the data on the propagation of top-down influences in freshwater pelagic ecosystems synthesized by McQueen *et al.* (1986) by plotting regressions similar to those in Figure 8.2(*a*). This was done by varying d_3, the mortality rate of the first-order carnivore, over a range of values, thus simulating a range of strengths of carnivory (higher carnivory being associated with lower values of d_3). The other parameter values were the same as before (Figure 8.3) except that two different nutrient inputs, I_n, were used, one an order of magnitude higher than the other. The resultant steady-state values N^*, N_x^*, N_y^* and N_z^*, as well as the total nutrient level, $N_T = N^* + N_x^* + N_y^* + N_z^*$, were computed for each value of d_3 in both the high and low nutrient input cases. (Remember, the variables N_x^*, N_y^*, N_z^* and N_w^* represent nutrient amounts, not biomass, in each of the trophic levels.) The pairs (N_x^*, N_T), (N_y^*, N_x^*) and (N_z^*, N_y^*) were plotted (Figure 8.4). The results show a striking negative correlation of the first-order carnivore with the herbivore and of the herbivore with the autotroph. On the other hand, the autotroph is strongly positively correlated with the total limiting nutrient. By comparing the higher nutrient input level (H) with the lower level (L) in the three frames of the figure, it can be seen that the level of nutrient input has a positive effect on all trophic levels (similar to Figure 8.3*b*).

8.3 EFFECTS OF HIGHER TROPHIC LEVELS ON NUTRIENT RECYCLING

Not only can nutrient input influence the structure of food webs but, reciprocally, food web structure can influence the level of available nutrient through recycling of nutrients by higher trophic levels. In the preceding chapters the role of herbivores in recycling nutrients back to the primary producers was discussed in detail. In particular, by 'short circuiting' the path of nutrients to the available pool and avoiding storage in detritus, where nutrients might become lost in sediments, regeneration by herbivores can keep nutrient molecules in the available pool for longer periods of time than they would be otherwise (e.g. Rigler, 1973).

Here we consider a further elaboration of this phenomenon that arises because of the presence of higher trophic levels. It has been hypothesized that a change in the turnover time of zooplankton could affect nutrient cycling (Henry, 1985). This could occur through the changing of the mean body size of zooplankton, since nutrient uptake and release rates are inversely proportional to body size (e.g. Peters and Rigler, 1973; Peters, 1975). Henry was able to reduce the mean size of zooplankton in plastic field enclosures by imposing artificial planktivory. He hypothesized that decreasing the mean body size of the zooplankton would increase the amount of phosphorus recycled within the epilimnion. In Chapter 6 it was shown that

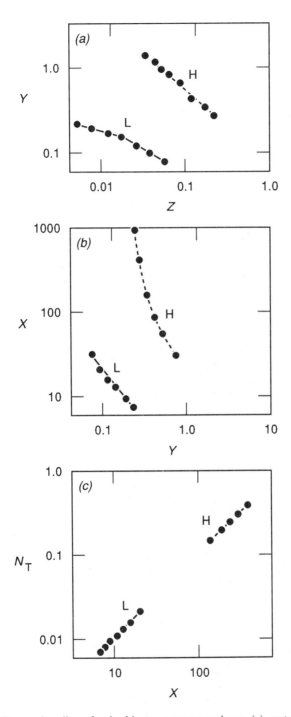

Figure 8.4 Regression lines for herbivore versus carnivore (*a*), autotroph versus herbivore (*b*) and total nutrient versus autotroph (*c*). Carnivore mortality, d_3, is varied to obtain the relationships and two values of I_n are used, 0.00005 (L) and 0.0005 (H). All other parameter values are the same as in Figure 8.3.

an increase in uptake rate of zooplankton could have this effect (Figure 6.8). Three different treatments were performed by Henry, (1) no artificial predator, (2) an artificial predator with a 200 µm filter and (3) an artificial predator with an 80 µm filter, thus decreasing the mean body sizes of remaining zooplankton in each case. Using phosphorus radioisotopes, Henry calculated the system turnover time (T_{res}) for P for each system (T_{res} being the time it took for phosphorus to reach the sediments at the bottom of the water column through diffusion and organism death and sinking). The filtration removal of larger-sized zooplankton increased T_{res} from 1.7 to 2.65 days, which led to increased availability of P; that is, it remained in the water volume for a longer time. Even small increases in availability could have significant effects (Bartell, 1981).

Mills *et al.* (1987) studied the cascading effects of fish predation on the Oneida lake food chain. They hypothesized that the shift in zooplankton towards smaller-bodied species, as young yellow perch suppressed the *Daphnia*, could influence nutrient cycling (in a slightly different way from the preceding case). In particular, the shift in zooplankton would decrease feeding rates on phytoplankton, which could lead to a dramatic increase in algal standing stocks and a consequent decrease in available phosphorus. Furthermore, exhaustion of available phosphorus would eventually cause a massive phytoplankton die-off, which would regenerate available phosphorus. In 1977, this hypothesized behaviour occurred. The *Daphnia pulex* population collapsed, total chlorophyll, which is proportional to algal production, peaked and the phosphorus level increased dramatically following a decline in algal abundance.

8.4 NUTRIENT LIMITATION AS A STABILIZING AGENT IN FOOD CHAINS

Local stability has long been a central issue for theoretical ecologists studying food webs. The ecologist is interested in not just the steady-state structures discussed in the preceding sections, but also in whether a given steady state is stable or whether it will collapse or oscillate wildly at the slightest perturbation. One of the basic discoveries in theoretical ecology during the 1970s was May's (1972, 1973) model result that large food webs tend to be unstable, with the probability of instability increasing as the number of components and interconnections between components increases. The inclusion of realistic constraints on energy flows between species in large food web models, however, was shown to decrease the chance of instability (Austin and Cook, 1974; DeAngelis, 1975). Another touchstone result of the 1970s was Rosenzweig's (1971) prediction of the 'paradox of enrichment', discussed earlier, which indicated that decreasing the level of nutrient in a system with a nutrient-limited autotroph population and a herbivore population decreased the probability of local instability of the system and thus of limit-cycle behaviour. Rosenzweig's results thus speak for a favourable effect of nutrient limitation on stabilizing food webs.

As reviewed in Chapter 5, many investigators have followed up on the idea that nutrient limitation can help stabilize food webs. It has been established that food web models of types commonly used, when adapted to represent nutrient-limited systems (either closed or open to the limiting nutrient) tend to be asymptotically stable. This is always true for linear donor-controlled systems such as Equations (8.1), as can be shown from the theorem introduced in the next section.

Even models of long food chains that have mildly destabilizing trophic interactions (or production functions) such as the Lotka–Volterra interaction, $F_2(N_x, N_y) = f_1 N_x N_y$, $F_3^*(N_y, N_z) = f_2 N_y N_z$ etc., can be stabilized by certain conditions on the nutrient cycling. Nisbet and Gurney (1976) and Nisbet et al. (1983) proved this for an n-level trophic chain with a closed nutrient cycle with Lotka–Volterra interactions between the trophic levels. The autotroph's growth rate, $F_1(N, N_x)$ in the model could be any function of the form $N_x U(N, N_x)$ as long as $dU/dN_x < 0$. As long as there were direct losses of nutrients from each trophic level back to the nutrient pool, the system always returned to the same steady state following any perturbation that left the total nutrient level, N_T, in the system unchanged (nutrient conserving perturbation). In the food web shown in Figure 8.1 all the nutrient goes from trophic levels into a detritus compartment before going back to the nutrient pool, but because the transition from detritus to the nutrient pool is donor-controlled, Nisbet and Gurney's proof would appear likely to apply to this model when all of the interaction terms, F, have the form of Lotka–Volterra functions.

The above result is strongly indicative that nutrient limitation in a closed system tends to counteract the tendency of trophic levels to interact in an oscillatory manner. Nisbet et al. (1983) concluded that the observed difficulties that experimentalists have in sustaining steady-state systems in microcosms and mesocosms over long time scales is not an inherent characteristic of closed systems in general. However, the authors were able to show through further investigations of the model that closed systems are highly sensitive to external perturbations (i.e. perturbations that violate the conditions of closedness of the system), which can easily provoke large fluctuations in the compartments.

8.5   RESILIENCE OF FOOD CHAINS

The resilience of a system, such as a population, nutrient in a lake, or a food web, was defined in Chapter 2 and was analysed in detail for autotroph and herbivore–autotroph systems in Chapters 3 and 5 respectively. Here we consider food web resilience more generally.

The resilience of mineral cycles has received attention in recent years.

Jordan *et al.* (1972) considered the stability of mineral cycles in forest ecosystems and proposed a standard model consisting of compartments for minerals in soil, wood, canopy leaves and litter. These compartments formed a cycle, but the system was an open one since minerals entered the system from rainfall and from the weathering of soil, and were lost via soil runoff and leaching. Data for various mineral cycles for three different forest types were used to parameterize the standard model, which was then run on the computer to obtain the recovery time, T_R, similar to that defined by Equation 2.7.

An important conclusion reached by Jordan *et al.* (1972) was that models of non-essential minerals tend to be more resilient than models of essential nutrients. The authors suggested, as a possible explanation for this peculiarity, that essential nutrients, such as calcium or phosphorus, tend to be tightly cycled. A perturbation to the system, therefore, damps away slowly. Minerals that are not essential, caesium for example, are lost at a high rate from the system, so the perturbation is quickly 'washed out'.

A similar observation was made by Pomeroy (1970), who pointed out that coral reefs and rain forests are examples of systems with tight nutrient cycles. As explained earlier in Chapter 3, when systems of this type are disturbed, by biomass removal causing significant nutrient loss, for instance, recovery may be very slow because there is little throughflow of nutrients coming from outside the system compared with the nutrients lost.

One can use a model of nutrient cycling in a food web to demonstrate the hypotheses of Jordan *et al.* (1972) and Pomeroy (1970) that tight cycling of nutrients results in increased time for recovery, T_R, or decreased resilience. Consider the set of Equations (8.1) for resource-controlled production of the trophic levels of a food chain. In Chapter 2 (e.g. Figure 2.7), it was shown that the return time, T_R, to equilibrium following a disturbance of a stable food web model could be approximated by the inverse of the real part of the dominant or critical eigenvalue; that is, the eigenvalue that has the smallest real part. [This approximation was, for example, used by Pimm and Lawton (1977) in their estimates of model food web resilience.]

If we initially make the assumption that there is no feedback or recycling of limiting nutrient from the detritus to the nutrient pool (that is, $d_D = 0$), then the eigenvalues of Equations (8.1) can be found almost by inspection, and are

$$\lambda_1 = -(r_1 + r_n) \qquad \lambda_2 = -(d_1 + e_1 + f_2)$$
$$\lambda_3 = -(d_2 + e_2 + f_3) \qquad \lambda_4 = -(d_3 + e_3 + f_4)$$
$$\lambda_5 = -(d_4 + e_4) \qquad \lambda_6 = -e_D$$

All of these eigenvalues are real and negative. The smallest eigenvalue in absolute value (closest to zero) dominates the long-term behaviour of the system. Suppose, for example, that λ_5 is the smallest eigenvalue. Then

$$T_R \cong 1/\lambda_5 = 1/(d_4 + e_4) \tag{8.4}$$

Note that the magnitudes of the six eigenvalues shown above are equal to the inverses of the nutrient turnover times of each compartment. For example, the turnover time of the herbivore compartment is $N_y^*/$(nutrient flux through compartment) or

$$N_y^*/[(d_2 + e_2 + f_3)N_y^*] = 1/(d_2 + e_2 + f_3)$$

This implies that longer food webs of this type will generally be lower in resilience (have longer turnover times) simply because there are more species in longer food chains than in shorter ones and thus a greater chance of there being a species population with a longer turnover time.

What happens when we change the situation by allowing tight recycling of the nutrients? This is accomplished mathematically by letting $d_D > 0$ and making

$$r_n, e_1, e_2, e_3, e_4, e_D \ll r_1, d_2, d_3, d_4, d_D \tag{8.5}$$

These inequalities mean that all of the fluxes of nutrients out of the system are small compared with the recycling fluxes. The average nutrient atom makes many circuits through the food chain before exiting. Now it is much harder to solve for the eigenvalues (it would require solving a sixth-order algebraic equation, which is impossible to do analytically, so that numerical solution would be necessary). However, one can establish bounds on the smallest eigenvalue by use of some theorems from mathematical matrix theory that are due to Perron, Frobenius and Collatz (see Varga, 1960, or Funderlic and Heath, 1971). The theorem applied to an 'essentially positive matrix', that is, one that has all off-diagonal elements positive, as is the case for the matrix associated with Equations (8.1):

$$\begin{vmatrix} -(r_1+r_n) & 0 & 0 & 0 & 0 & d_D \\ 0 & -(d_1+e_1+f_2) & 0 & 0 & 0 & 0 \\ 0 & \eta f_2 & -(d_2+e_2+f_3) & 0 & 0 & 0 \\ 0 & 0 & \eta f_3 & -(d_3+e_3+f_4) & 0 & 0 \\ 0 & 0 & 0 & \eta f_4 & -(d_4+e_4) & 0 \\ r_1 & d_1+(1-\eta)f_2 & d_2+(1-\eta)f_3 & d_3+(1-\eta)f_4 & d_4 & -(d_D+e_D) \end{vmatrix} \tag{8.6}$$

A matrix of this type has one real eigenvalue δ that has a smaller absolute value than any other real part of an eigenvalue of the system. Also, either all of the column sums of (8.6) equal the eigenvalue δ or δ lies between the maximum and minimum of the column sums of the matrix.

To put bounds on δ, the eigenvalue that governs the long-term behaviour of the model system, it is only necessary to find the column sums of the matrix (8.6). These are: (1) $-r_n$; (2) $-e_1$; (3) $-e_2$; (4) $-e_3$; (5) $-e_4$; (6) $-e_D$. Suppose that one of these column sums, r_n for example, is larger in absolute value than any of the others. Then the absolute value of the smallest eigenvalue, δ, is, by the above theorem, smaller than r_n; $|\delta| < r_n$.

This result enables us to compare the resilience of this system with the one in which there was no nutrient recycling. Recall that the return time T_R for the system with no recycling was given approximately by Equation (8.4), or

$$T_R = 1/(d_4 + e_4)$$

whereas T_R for the system with recycling is greater than $1/r_n$. From inequality (8.5), $r_n \ll d_4$, so that

$$1/(d_4 + e_4) \ll 1/|\delta| \tag{8.7}$$

that is, the return time for the food chain model with recycling is much greater and the resilience much less than the food chain model without recycling.

It is possible to show also that when inequality (8.5) holds, the nutrient turnover time T_{res} is of roughly the same magnitude as T_R (i.e. $> 1/r_n$). To demonstrate this, note that when the inequality is satisfied, the total nutrient

$$N_T^* \doteq N^* + N_x^* + N_y^* + N_z^* + N_w^*$$

scales as I_n/r_n. This occurs because in the limit in which the terms on the left-hand side of inequality (8.5) approach zero relative to the terms on the right-hand side, K in Equation (8.2a) approaches r_1 (as I will leave for the reader to show). Then the nutrient turnover time for the system is

$$T_{res} = N_T^*/I_n = (1/r_n)[1 + r_1/(f_2 + d_1) + \cdots]$$

where the additional terms within the brackets can be found from Equations (8.2c,d,e). Thus, $T_{res} > 1/r_n$ and can be used to estimate roughly the magnitude of T_R.

In the above demonstration, a linear donor-controlled food chain was chosen for simplicity, but the argument can be extended to non-linear food chains and even to food webs (see DeAngelis, 1980).

The above results point to an important generalization relating resilience and nutrient flow through food webs. When nutrient input is the limiting rate process, the resilience of a food web can be represented by the ratio of the steady-state input or loss rate (the two being equal) of the nutrient, I_n, to the steady-state stock of nutrient in the system, N_T. The higher I_n/N_T is (which usually means a higher nutrient subsidy to the system), the greater is the resilience of the food web to perturbations that change the level of limiting nutrient in the system.

8.6 DO ECOSYSTEMS CHANGE TOWARDS TIGHTER NUTRIENT CYCLING?

Because nutrient cycling has important consequences for food web resilience, as shown above, it is important to consider how nutrient cycles develop in ecosystems during the process of ecological succession. The process of succession has been defined by Odum (1969, 1971) as an orderly and reasonably predictable process of community development that involves changes in species structure and community processes through time, culminating in a 'stabilized ecosystem in which maximum biomass (or high information content) and symbiotic function between organisms are maintained per unit of available energy flow'. Odum identified the '... closing or 'tightening' of the biogeochemical cycle of major nutrients, such as nitrogen, phosphorus, and calcium...' as an important successional trend. This increased tightening of nutrient cycles would seem to imply a trend towards smaller nutrient losses from the system as succession proceeds.

Actually, this concept is not quite as simple as it first seems. Does this tightening of nutrient cycling mean a reduction in the absolute loss rate, the ratio of output to input (O_n/I_n), the ratio of output to standing stock (O_n/N_T), the ratio of output to recycled flux (O_n/TST_c) [Equation (3.16) of Chapter 3] or an increase in some cycling index, such as Finn's cycling index, CI [Equation (3.16) of Chapter 3]? Finn (1982) discussed Odum's generalization in terms of its implications for all four of these indices, which are shown schematically in Figure 8.5. Note that Odum's concept does not make any predictions about changes in the ratio of output to input through successional time.

Some ecologists have disagreed with the homoeostatic view and, in

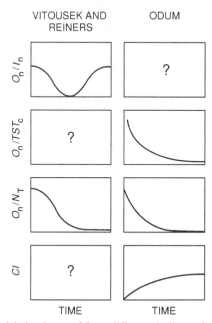

Figure 8.5 Predicted behaviours of four different indices of nutrient output based on the conceptual models of Odum (1969, 1971) and Vitousek and Reiners (1975). (From Finn, 1982.)

describing succession, have attributed far less regularity and determinism to the patterns of species replacement than this view; see, for example, Drury and Nisbet (1973) and Shugart (1984). The homoeostatic view as applied to nutrient cycling (that is, the view that there is a trend towards tighter cycling during succession) has also been questioned.

Vitousek and Reiners (1975) presented a conceptual model for nutrient relationships in a successional ecosystem that is an alternative to Odum's (1969, 1971). In their view the nutrient output rate in the early successional stages is high, then drops to a very low value during the intermediate stages, and finally returns to higher values to approximately balance nutrient inputs in later successional stages. The early nutrient output production would vary in detail depending on whether the succession is 'primary' (starting on bare substrate) or 'secondary' (succession following a disturbance on an existing ecosystem).

This general pattern of nutrient flux through the system is influenced by the successional pattern of net ecosystem production. Net biomass production starts slowly in primary succession. In secondary succession this production may even drop below zero (net loss of biomass) in early stages. The production rate builds up to a peak during the intermediate period of succession and then declines to a lower steady-state value when the biomass density of the system asymptotes in late successional stages. Nutrient accumulation is highest during the peak of net ecosystem biomass production because nutrient is stored in the increasing standing stock of biomass. Nutrient output can be calculated from a balance of nutrient fluxes:

Nutrient output = Nutrient input − Nutrient accumulation   (8.8)

If the nutrient input is roughly constant, then a characteristic trough-like pattern in the output rate is formed over successional time, in which the ratio of nutrient output to input (O_n/I_n) actually increases in later successional stages.

Vitousek and Reiners' conceptual model is shown in Figure 8.5. Their model also predicts a decrease in (O_n/N_T), though in a more sigmoidal form than Odum's (1969, 1971). However, Vitousek and Reiners' model makes no predictions concerning (O_n/TST_c) or CI.

Finn's (1982) comparison shows that there is no necessary inconsistency between the conceptual models of Odum (1969, 1971) and Vitousek and Reiners (1975). In fact, Finn tested these conceptual ideas by means of a simple simulation model of a hypothetical successional sequence. This model included an external nutrient input, growth of ecosystem biomass up to a carrying capacity and the capacity for recycling of nutrients. Finn simulated succession with the model and computed (O_n/I_n) and CI through time for two cases: high recycling and low recycling. In Figure 8.6(a), the trough-like (O_n/I_n) ratio predicted by Vitousek and Reiners is clearly indicated for both cases. Figure 8.6(b) shows that when recycling is high, CI steadily increases at the same time that (O_n/I_n) undergoes trough-like

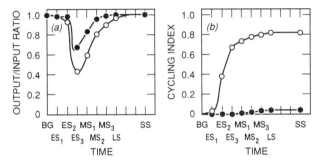

Figure 8.6 Calculations of the ratio of nutrient input to output (O_n/I_n) and of the cycling index *CI* based on a simulation model of succession. The horizontal axis represents successional time. The two cases are high recycling (open circles) and low recycling (closed circles). (From Finn, 1982.)

behaviour. Thus, there is at least no inconsistency between the hypothesis of Odum and that of Vitousek and Reiners. Of course, these hypotheses must be tested with data and as yet few data of this type exist over successional time periods.

The details of these patterns of indices of nutrient output depend on specific mechanisms that may influence particular nutrients in particular ecosystems. These mechanisms include: input mechanisms such as weathering, nitrogen fixation, dry and wet deposition; output mechanisms such as leaching and erosion; and internal mechanisms such as net ecosystem production, decomposition and element mobilization, regulation of soil solution chemistry and variability in the utilization of elements by the biota (Gorham *et al.*, 1979). Nutrients that are essential to the biota may be retained within the system by various mechanisms that slow their rate of loss. The authors make the point that examination of these mechanisms in detail is indispensable in determining the pattern of nutrient output from the system.

8.7 NUTRIENTS AND CATASTROPHIC BEHAVIOUR IN ECOSYSTEMS

Earlier in this chapter, we saw examples of a model in which increases in nutrient input caused gradual continuous changes in the higher trophic levels. This smoothness of food web response to changes need not always be the case. In Chapter 4 we analysed in detail how catastrophic behaviours can emerge in food chains under specific assumptions about interactions. In particular, slow changes in nutrient input could lead to sudden drastic changes in the state of a forest ecosystem (Gatto and Rinaldi, 1987). Ecological systems are prone to such instabilities, which are often called structural instabilities (e.g. Ludwig *et al.*, 1978). One example of such a catastrophe occurred in a model by Voinov and Svirizhev (1984) of the eutrophication process in a system of primary producers, available nutrient, detritus and oxygen. The total amount of limiting nutrient in the system was gradually increased. The assumption was made that the mortality and aerobic decomposition rate of the phytoplankton exceeded its production of oxygen. As the total nutrient level N_T increased, a point was reached at which oxygen diminished to zero, so that the entire system changed to an anaerobic one. This prediction is corroborated in many situations in which high concentrations of nutrient occur in bodies of water. Frequently, the change is amplified by positive feedback mechanisms. For example, anoxia occurs in the deeper waters of the central portions of the Chesapeake Bay as a result of stratification and the respiration of dead organic matter that sinks down from the euphotic zone. Officer et al. (1984) suggested that conditions of anoxia in bottom waters, by causing more rapid mineraliz- ation of phosphorus from the sediments, may stimulate increased primary production (Mortimer, 1942). If the condition holds that phytoplankton mortality exceeds its oxygen production, this would shift the system towards increased anoxia.

A somewhat different means by which a complex food web may be pushed from one state to another is through large perturbations, such as pulses of nutrients. Neill (1988) experimentally explored the stability of food webs in oligotrophic montane lakes. The lakes studied were generally dominated by herbivorous zooplankton. A possible predator, the phantom midge *Chaoborus trivittatus*, usually did not flourish well enough to exert control on the zooplankton. It died off at a high enough rate that few individuals reached large enough sizes to feed on the *Daphnia*. However, a large pulse of nutrients, particularly when accompanied by experimental removal of some of the zooplankters (*Daphnia*), resulted in a dramatic change in community structure, creating one dominated by *Chaoborus*. The dramatic shift resulted from greater primary production and from an increase in rotifers, a food source of first instar *Chaoborus*, caused by reduced predation on rotifers by *Daphnia*. The enhanced *Chaoborus* survival allowed large later instars to survive and to further suppress *Daphnia*. This positive feedback effect led to large-scale changes in community structure. Although these were not permanent changes, they at least suggest a stable alternative state.

This type of behaviour may be characteristic of large complex systems, in which a sufficiently large perturbation in nutrient levels can lead to changes in organism behaviours and survival rates.

8.8 SUMMARY AND CONCLUSIONS

Food webs in nature are complex and may in general contain trophic chains of up to four or five levels. The way in which effects propagate up such chains from the resources to the top and from top consumers down to the bottom are important properties of such chains. Both empirical evidence and mathematical models indicate that effects propagate in both directions. Resource inputs in part determine the number of trophic levels and the amount of biomass that can be built up in these levels, though the effects of increases in resource input may eventually decline because of saturation. Top-down effects, such as trophic interactions that depend on consumer population biomasses, may have effects down to the autotroph level.

Higher trophic levels, such as herbivores and carnivores, can have a significant effect on nutrient recycling, decreasing nutrient turnover time in the food web by creating short circuits of nutrients directly back to the nutrient pool without lengthy delays in the detritus.

Recycling affects food chain resilience. Systems with a high degree of recycling are generally less resilient that those that allow nutrients to flow through without much recycling. The latter systems have a much higher turnover rate of nutrients and can, as far as nutrient needs are concerned, restore biomass losses more quickly. Essential nutrients are often recycled to a high degree in a system. This allows a higher rate of biomass production in the system, but renders it vulnerable to disturbances that remove biomass.

Complex food webs with particular kinds of interactions are prone to catastrophic changes from one state (ecological community) to another. Such phenomena may occur as sudden changes that result during slow changes in some parameter, such as nutrient input. Large pulses of nutrients may also cause quasi-permanent changes in the ecological community.

9 Competition and nutrients

9.1 INTRODUCTION

Food webs are characterized not only by the lengths of the chains linking nutrient and energy flows, but also by the total number of species in the web and the number of species at each trophic level. Community diversity is to some extent a measure of the number of species that fill similar functional roles. Species occupying the same trophic level may compete for resources represented by lower trophic levels. The nature of the relation between species and their resources (energy and nutrients from the lower trophic levels) determines how many species populations may share a trophic level in a particular environment.

Competition has been defined to occur 'whenever a valuable or necessary resource is sought together by a number of animals or plants (of the same kind or of different kinds) when that resource is in short supply; or if the resource is not in short supply, competition occurs when the animals or plants seeking that resource nevertheless harm one another in the process' (Andrewartha and Birch, 1954).

Competition between different species, called interspecific competition, has been investigated by means of laboratory experiments (early references include Gause, 1934; Crombie, 1947; Park, 1954). When similar species are placed together under controlled experimental conditions, the one species that is best at exploiting the particular experimental environments is usually able to eliminate the others. This phenomenon has been termed 'competitive exclusion' and has been termed 'competitive exclusion' and has been developed as a mathematical theory, as will be described later. Field studies, including experimental manipulation (e.g. Connell, 1961; Inouye, 1978), have also been performed that demonstrate competitive effects; that is, one species increases in number when the other is removed. This implies that the species share one or more resources in common: substrate in the intertidal area in the case of barnacles studied by Connell and nectar of some of the same flowers in the case of the bumblebees studied by Inouye. However, under natural conditions the species do not competitively exclude each other because each has a habitat or a resource for which it is better adapted and specialized than the other species.

Plants generally compete for light, water and nutrients, which is a smaller and more homogeneous set of resources compared with the wide array of different resources available to animals (Grubb, 1977; Connell, 1978; Tilman,

1982). This difference between the two organism types has made explaining the observed diversity of plants in food webs more difficult than that of explaining the diversity of animals. Animal diversity can easily be interpreted in terms of different animal species specializing on different resources, whereas the possibility for specialization is not obvious for plants. However, a number of interesting hypotheses for plant diversity have been advanced.

In this chapter, competition by autotrophs for nutrients is examined and the effects of this competition on plant diversity and plant succession are discussed. The interaction of higher trophic levels with autotroph nutrient competition is also considered.

9.2 EXPLOITATIVE COMPETITION FOR NUTRIENTS

Exploitative competition is an indirect form of competition in which each individual that shares a resource lowers the ability of other individuals to survive and reproduce by decreasing the amount of the resource available to the others. This competition for resources can occur among individuals within a species and also between individuals of two or more different species when their resources are similar enough. It is the occurrence of interspecific competition that is of interest here. This can lead to competitive exclusion if the growth rates of all species depend on a single resource. Consider the simple model for two autotrophs, utilizing a single nutrient of amount N in a system in which the growth rates of both are governed by Monod functions:

$$dN/dt = I_n - r_n N - [r_1 N_{X1} N/(k_1 + N)] - r_2 N_{X2} N/(k_2 + N) \qquad (9.1a)$$

$$dN_{X1}/dt = [r_1 N_{X1} N/(k_1 + N)] - d_1 N_{X1} \qquad (9.1b)$$

$$dN_{X2}/dt = [r_2 N_{X2} N/(k_2 + N)] - d_2 N_{X2} \qquad (9.1c)$$

Here N_{X1} and N_{X2} are amounts of nutrient in the biomass of each autotroph (equivalent nutrient amounts). Nutrient recycling is ignored here for simplicity.

At equilibrium, Equation (9.1b) implies that

$$N^* = k_1 d_1/(r_1 - d_1) \qquad (9.2a)$$

but Equation (9.1c) implies that

$$N^* = k_2 d_2/(r_2 - d_2) \qquad (9.2b)$$

The two equations are inconsistent. If

$$k_1 d_1/(r_1 - d_1) < k_2 d_2/(r_2 - d_2) \qquad (9.3)$$

then N_{X2} will go to extinction. This occurs because autotroph 1 lowers the steady-state nutrient level, N^*, to $k_1 d_1/(r_1 - d_1)$, at which level the inequality $dN_{X2}/dt < 0$ occurs. When the model is generalized so that n autotrophs compete for this same resource, only one autotroph will survive (O'Brien, 1974; Hsu *et al.*, 1977; Armstrong and McGehee, 1980; Tilman, 1982). If the inequality (9.3) were reversed, then N_{X1} would go to extinction and N_{X2} would survive. If (9.3) were an equality then, in principle, the two species could coexist. However, the equality would have to be very precise, which is unlikely in nature.

Abrams (1988) has criticized the conclusion from this simple model that one limiting resource cannot support two or more competing species on the grounds that even a presumably simple nutrient resource such as nitrogen or phosphorus actually exists in many forms, depending on soil structure, climate and other factors; thus it constitutes many 'resources'. Nonetheless, the result is a useful starting point for consideration of factors that promote coexistence of competitors.

Tilman (1982) presented some examples of experiments illustrating competitive exclusion on the basis of the above criterion. What is interesting is that criterion (9.3) allows a prediction to be made concerning which species will exclude which other species based on individual measurements of the growth rates of the species. Tilman *et al.* (1981) studied two diatom species (*Asterionella formosa* and *Synedra ulna*) taken from Lake Michigan and measured their growth rates as functions of the limiting nutrient, silicate. The authors showed that *Synedra* alone reduced nutrient concentration further than did *Asterionella* alone. When allowed to compete, *Synedra* displaced *Asterionella*, consistent with the criterion (9.3). A number of other experiments, including tests on terrestrial plants (e.g. Braakhekke, 1980) have shown the same patterns of competitive exclusion.

The system described by Equations (9.1a,b,c) displays competitive exclusion because the autotrophs control the equilibrium nutrient level, N^*. It is also possible for nutrient level to mediate competition, even when the competition does not involve exploitation of the nutrient resource. Consider the Gause model for competion between two species, 1 and 2:

$$dX_1/dt = r_1(1 - X_1/K_1 - \alpha_{12}X_2/K_1)X_1 \tag{9.4a}$$

$$dX_2/dt = r_2(1 - X_2/K_2 - \alpha_{21}X_1/K_2)X_2 \tag{9.4b}$$

where X_1 and X_2 are biomasses of the two competitors here. This model differs from that of Equations (9.1a–c) in the respect that the resource is not explicitly represented here. The competitors are not assumed to have any effect on the resource level here, as they do in the model described by Equation (9.1a–c). Under the conditions $K_1/\alpha_{12} > K_2$ and $K_2/\alpha_{21} > K_1$, there can be stable coexistence between these species at the equilibrium point E. This is shown in Figure 9.1(*a*), where the zero isocline is $dX_1/dt = 0$ and $dX_2/dt = 0$ are plotted. When the signs of both of these inequalities are reversed, there is an equilibrium point, E, at which both species exist, but the equilibrium is unstable and either species may be displaced, depending

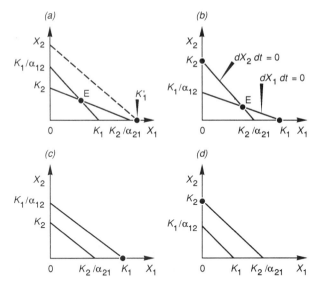

Figure 9.1   Phase-plane diagrams showing the possible ways in which two competing species populations may interact, when the populations are described by Lotka–Volterra equations. The two lines are the zero isoclines $dX_1/dt = 0$ and $dX_2/dt = 0$. The point E in (*a*) is the only equilibrium point for which the two competing species coexist stably. The dashed line in (*a*) represents the effect of an increased nutrient level that increases the carrying capacity only of species 1. Equilibrium point E disappears and is replaced by a new equilibrium point along the X_1 axis.

on initial conditions (Figure 9.1*b*). For example, a perturbation away from E that decreases X_2 only will cause X_2 to go to zero and X_1 to go towards K_1. The other possible situations are more deterministic. If $K_1/\alpha_{12} > K_2$ but $K_2/\alpha_{21} < K_1$, then species 1 will always displace species 2, whereas, when $K_1/\alpha_{12} < K_2$ and $K_2/\alpha_{21} > K_1$, species 2 will always displace species 1 (Figures 9.1*c,d*).

Suppose that the carrying capacities, K_1 and K_2, for one or both species are affected by the nutrient level, N, but the autotrophs do not affect the level of N. Under these circumstances, Riebesell (1974) showed theoretically that an increase in nutrient concentration can possibly destabilize an initially stable system, such as that of Figure 9.1(*a*). Suppose that K_1 increases with an increase in N, to a new value K'_1, whereas K_2 is relatively unaffected. Then, as N increases, the $dX_1/dt = 0$ isocline moves to the right (see dashed line in Figure 9.1*a*) to the point where E is eliminated. The new equilibrium point is $X_1^* = K'_1$, $X_2^* = 0$.

Riebesell (1974) has called this effect the 'paradox of enrichment in competitive systems' in analogy with the 'paradox of enrichment in predator–prey systems' that Rosenzweig (1971) introduced (see Chapter 5). The lesson is the same: an increase in the nutrient resource level in the system may have a destabilizing effect on the system.

9.3 EFFECTS OF HIGHER TROPHIC LEVELS

The fact that competitive exclusion necessarily occurs in the model described by Equations (9.1) is a result of the autotrophs having a controlling effect on nutrient level, N. When herbivores are present, the situation is quite different and the autotrophs may not be able to control the nutrient level. Suppose that each autotroph has an herbivore specialized on it, so that the feeding rates described by Holling Type III, $f_1 N_{x1}^2 N_{y1}/(b_1 + N_{x1}^2)$ and $f_2 N_{x2}^2 N_{y2}/(b_2 + N_{x2}^2)$ are subtracted from Equations (9.1b) and (9.1c) respectively and two additional equations are added to describe the nutrient equivalent amounts, N_{y1} and N_{y2}, in the two herbivores:

$$dN_{y1}/dt = f_1 N_{x1}^2 N_{y1}/(b_1 + N_{x1}^2) - d_3 N_{y1} \tag{9.1d}$$

$$dN_{y2}/dt = f_2 N_{x2}^2 N_{y2}/(b_2 + N_{x2}^2) - d_4 N_{y2} \tag{9.1e}$$

(Figure 9.2a). The solutions of the $dN_{y1}/dt = 0$ and $dN_{y2}/dt = 0$ imply that the autotrophs are held to the steady-state levels

$$N_{x1}^* = [b_1 d_3/(f_1 - d_3)]^{1/2} \tag{9.5a}$$

$$N_{x2}^* = [b_2 d_4/(f_2 - d_4)]^{1/2} \tag{9.5b}$$

The steady-state value of nutrient, N^*, is found by setting $dN/dt = 0$ and solving the right-hand side of Equation (9.1a) (it is a second-order algebraic equation that will be left to the reader), and N_{y1}^* and N_{y2}^* are given from solutions of (9.1b) and (9.1c) (remember that a consumption term is assumed to each equation);

$$N_{y1}^* = [r_1 N^*/(k_1 - N^*) - d_1](b_1 + N_{x1}^{*2})/(f_1 N_{x1}^{*2}) \tag{9.6a}$$

$$N_{y2}^* = [r_2 N^*/(k_2 + N^*) - d_2](b_2 + N_{x2}^{*2})/(f_2 N_{x1}^{*2}) \tag{9.6b}$$

For appropriate parameter values, both N_{y1}^* and N_{y2}^* are positive and an equilibrium point occurs where all four species coexist.

Levine (1976) pointed out an interesting characteristic of systems like that in Figure 9.2(a). Suppose the steady-state equilibrium level of one of the herbivores, N_{y1}^*, is increased, by decreasing the mortality rate of that herbivore population, d_3, for example. Levine showed that the effect of the increase in N_{y1}^* is to increase the equilibrium value of the other herbivore, N_{y2}^*. This result at first appears to be counterintuitive, because the $N_{x1} - N_{y1}$ and $N_{x2} - N_{y2}$ chains are competing for one resource, N. However, if one examines the chain of connections between N_{y1} and N_{y2}, it is clear that the species N_{y1} and N_{y2} do not actually compete and, in fact, behave in a way that resembles mutualism (though it is not a true mutualism). An increase in steady-state N_{y1}^* causes N_{x1}^* to decrease, which

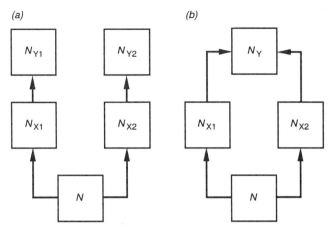

Figure 9.2 (a) Two autotrophs competing for the same nutrient, N, and grazed on by specialized herbivores with biomasses Y_1 and Y_2 (and corresponding nutrient amounts N_{y1} and N_{y2}); (b) two autotrophs competing for a single limiting resource and grazed on by a single generalist herbivore.

increases the level of nutrients, N^*. This has the effect of increasing N_{y2}^*. Levine demonstrated this effect first in the model and Vandermeer (1980) specified more systematically the conditions under which the effect could occur.

The existence of food web configurations of the form shown in Figure 9.2 may not be uncommon in ecosystems. Kerfoot and DeMott (1984) presented some examples in aquatic ecosystems. They referred to the food chains that diverge from a common resource base as 'subordinate linkage hierarchies' and the vertical development of such parallel chains as 'vaulting'.

A special case exists when only one of the specialist herbivores is present. A three-species system can only be established if the herbivore feeds on autotroph 1, which is competitively superior in the absence of herbivory. Then, although autotroph 2, which has no herbivore feeding on it, controls the nutrient level, fixing it at N^* [Equation (9.2a)], N^* may be high enough to support the other food chain consisting of autotroph 1 and herbivore 1.

Another simple alternative food web configuration exists in which only one herbivore feeds on both autotrophs (Figure 9.2b). A possible set of equations for this system is

$$dN/dt = I_n - r_n N - [r_1 N_{x1} N/(k_1 + N)] - r_2 N_{x2} N/(k_2 + N) \qquad (9.7a)$$

$$dN_{x1}/dt = r_1 N_{x1} N/(k_1 + N) - [f_1 N_{x1} N_y/(b_1 + N_{x1})] - d_1 N_{x1} \qquad (9.7b)$$

$$dN_{x2}/dt = r_2 N_{x2} N/(k_2 + N) - f_2 N_{x2} N_y/(b_2 + N_{x2}) - d_2 N_{x2} \qquad (9.7c)$$

$$dN_y/dt = f_1 N_{x1} N_y/(b_1 + N_{x1}) + f_2 N_{x2} N_y/(b_2 + N_{x2}) - d_3 N_y \qquad (9.7d)$$

Steady-state analysis of this model is too complex to be done here in a limited space, but a simple trick can be used to gain information on whether the two autotrophs and herbivore can coexist. The trick consists in first letting one of the autotrophs be absent from the system, finding the steady-state equilibrium values of the remaining components, and seeing if the autotroph initially absent could invade (that is, if it would tend to increase in the system if a very small amount of it were added to the steady-state system). This process is then repeated with the other autotroph initially assumed to be absent.

Assume first, for example, that $N_{x2}^* = 0$. Then the other components have the equilibrium values

$$N^* = -(1/2)\{k_1 + [(r_1/r_n)N_{x1}^*] - I_n/r_n\} + (1/2)\{(k_1 + [(r_1/r_n)N_{x1}^*]$$
$$- I_n/r_n)^2 + 4(k_1/r_n)I_n\}^{1/2} \tag{9.8a}$$

$$N_{x1}^* = b_1 d_3/(f_1 - d_3) \tag{9.8b}$$

$$N_y^* = \{[r_1 N^*/(k_1 + N^*)] - d_1\}(b_1 + N_{x1}^*)/f_1 \tag{9.8c}$$

If, given these equilibrium values, the inequality, based on Equation (9.7c) for N_{x2} very small

$$(1/N_{x2})dN_{x2}/dt = [r_2 N^*/(k_2 + N^*)] - [f_2 N_y^*/b_2] - d_2 > 0 \tag{9.9}$$

holds, then an initially small value of N_{x2} can increase when autotroph 1 is present in steady state, so autotroph 2 can successfully invade. If the same is true for autotroph 1 when it is initially absent and autotroph 2 is in steady state, this strongly suggests that the species can coexist at equilibrium in the system. Various relative feeding rates of N_y on N_{x1} and N_{x2} can accomplish this.

A number of experiments have examined the effects of herbivores on the structure of a community of autotrophs competing for nutrients. These experiments are often performed by manipulation of nutrient levels and by either direct manipulation of the herbivore community (additions or removals of herbivores) or by indirect manipulation by addition or removal of predators of the herbivores.

As discussed in Chapter 5, when the herbivore is the top trophic level, herbivory tends to limit the autotroph level. It also tends to shift the competitive balance among autotrophs. For example, Vanni (1987) studied the phytoplankton community of a small temperate oligotrophic lake. Using enclosures with various nitrogen and phosphorus enrichment treatments and with fish present or absent (creating low and high herbivorous zooplankton levels respectively), Vanni showed that the proportion of phytoplankton density consisting of small-bodied edible species was lower in the fishless enclosures under all nutrient treatments.

Kerfoot et al. (1988) demonstrated that shifts in Daphnia densities in a small mesotrophic lake resulted in rapid shifts in algal assemblages (between naked flagellates and species more resistant to herbivory) though total phytoplankton cell densities stayed relatively constant.

Cukor (1983) studied the effects of nutrient enrichment and grazing by snails (*Lymnaea*) on the epilithic community of an Arctic lake. Both changes in nutrient loading and in grazing caused shifts in the algal community. *Lymnaea* grazing reduced large diatoms in relation to less accessible small green and blue–green algae.

Terrestrial systems show similar results. Berendse (1985) analysed information on two plant species in wet heathlands in the Netherlands: *Erica tetralix* and *Molinia caerulea*. *Erica* has a lower nutrient requirement than *Molinia*, whereas the latter is a better competitor under the existing rich nutrient conditions. However, sheep prefer to graze on *Molinia* and thus have traditionally kept this species from becoming dominant. Reduction in sheep grazing in recent years has led to *Molinia* increasing rapidly in relation to *Erica*. Berendse showed, using a model, that reintroduction of grazing may not be sufficient to shift a pasture away from a stable state of *Molinia* domination.

Reinertsen *et al.* (1986) showed that higher trophic levels may affect autotroph community structure in other ways than through grazing alone. Fish and zooplankton can also influence phytoplankton community structure by the way in which they recycle nutrients. Zooplankton release phosphorus relatively homogeneously over the spatial volume that they occupy in a body of water, whereas phosphorus release by fish is much more spatially patchy. Thus, in communities with fish populations, phytoplankton species such as *Anabaena flos-aquae*, which are capable of rapid uptake of temporary pulses of phosphate, can coexist with the algal species *Staurastrum luetkemuelleri*, whereas *Anabaena* is at a disadvantage when only zooplankton are present.

Higher trophic level communities themselves are affected by nutrient enrichment, responding to changes in phytoplankton communities. Patalas (1972) noted a gradient in the composition of zooplankton communities in the Great Lakes, including a decrease in calanoids and an increase in cyclopoids and cladocerans from oligotrophic Lake Superior to eutrophic Lake Erie.   .

9.4 SPECIES REPLACEMENT WITH INCREASING NUTRIENT LEVELS

The term 'succession' refers to changes in the species composition of an ecological community through time, usually following a disturbance that removes the current species populations. This succession of species populations may be a result of one or a combination of the following general factors: (1) phenotypic characteristics of species (some entering a disturbed area sooner than others and growing faster); (2) externally imposed changes in one or more environmental parameters that favour some species over others; (3) changes in the environment caused by the populations themselves (e.g. greater shading as biomass levels increase).

Armstrong (1979) modelled the process of succession under gradual changes in environmental conditions, namely, increases in nutrient level in the ecosystem, or 'eutrophication'. Armstrong used a model that, when applied to only two autotroph species, has a form similar to that of Equations (9.7). Armstrong's model differs from Equations (9.7) only in that his functional forms for trophic interaction are more general. With my notation for available nutrient, equivalent autotroph nutrient, N_{x1} and N_{x2}, and equivalent herbivore nutrient, N_y, Armstrong's model can be written as,

$$dN/dt = N_y w_s(N_{x1}, N_{x2}) - N_{x1} g_1(N) - N_{x2} g_2(N) \tag{9.10a}$$

$$dN_{x1}/dt = N_{x1}[g_1(N) - N_y f_1 h(N_{x1}, N_{x2})] \tag{9.10b}$$

$$dN_{x2}/dt = N_{x2}[g_2(N) - N_y f_2 h(N_{x1}, N_{x2})] \tag{9.10c}$$

$$dN_y/dt = N_y[f_1 h(N_{x1}, N_{x2}) + f_2 h(N_{x1}, N_{x2}) - w_s(N_{x1}, N_{x2})] \tag{9.10d}$$

Here, $w_s(N_{x1}, N_{x2})$ is a rate of nutrient regeneration by the herbivore, $g_1(N)$ and $g_2(N)$ are growth rates of the autotrophs and $h(N_{x1}, N_{x2})$ is the herbivore feeding rate. These functions are assigned general properties but not given explicit forms. The symbols f_1 and f_2 are constants representing the relative harvest rates of the two autotrophs.

Note that Armstrong's model is closed to nutrient flux. The total nutrient in the system is N_T:

$$N_T = N + N_{x1} + N_{x2} + N_y \tag{9.10e}$$

This equation can be used to replace Equation (9.10d) and thus simplify studying the system of equations. The total nutrient in the system, N_T, provides an unambiguous index of the level of eutrophication in the system. Another simplifying assumption is that the functional response, $h(N_{x1}, N_{x2})$, of the herbivore is related to the proportions that the two autotrophs contribute to the herbivore's diet, i.e.

$$h(N_{x1}, N_{x2}) = h(f_1 N_{x1} + f_2 N_{x2})$$

Armstrong presented a graphical means for analysing this system. First, he used Equation (9.10e) in Equations (9.10b,c) so that $g_1(N)$ and $g_2(N)$ were converted to $g_1(N_T - N_{x1} - N_{x2} - N_y)$ and $g_2(N_T - N_{x1} - N_{x2} - N_y)$. The functional forms of the growth rates, g_1 and g_2, are such that when g_1/f_1 and g_2/f_2 are plotted against $N_{x1} + N_{x2} + N_y$, curves of the form shown in Figure 9.3 are obtained. An increase in total nutrient in biomass, $N_{x1} + N_{x2} + N_y$, is the same thing as a decrease in available nutrient, N; thus, the autotroph growth rates, g_1 and g_2, should decrease as $N_{x1} + N_{x2} + N_y$ increases, as shown by the qualitative curves shown in Figure 9.3. If there is no herbivory ($N_y = 0$), then the point P_1' represents the level that N_{x1} of autotroph 1 reaches in the absence of autotroph 2 ($N_{x2} = 0$), since at this point the growth rate of autotroph 1 goes to zero. P_2' represents the level that N_{x2} of autotroph 2 reaches in the absence of autotroph 1 ($N_{x1} = 0$).

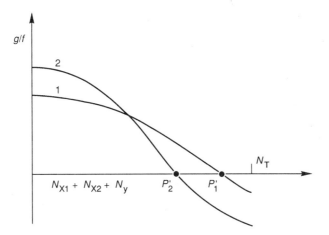

Figure 9.3 Normalized growth curves, g_1/f_1 and g_2/f_2, of two autotrophs utilizing a single limiting nutrient, N. The points $P_{1'}$ and $P_{2'}$ represent the equilibria of autotrophs 1 and 2, respectively, in the absence of the other autotroph. Note that autotroph 2 would be excluded in the absence of herbivory because it has a negative growth rate when autotroph 1 is present at equilibrium.

Now, suppose that autotroph 1 exists at steady-state equilibrium (P_1', where $g_1/f_1 = 0$, so that $dN_{x1}/dt = 0$), with autotroph 2 absent. If autotroph 2 were introduced into the system, it could not persist, because it would have a negative growth rate ($g_2/f_2 < 0$) since the curve g_2/f_2 is below the zero axis at that level of available nutrient, $N = N_T - N_{x1}$. Therefore, autotroph 2 would be excluded in the absence of herbivory.

When the herbivore is added and one autotroph is assumed to be present, the effects of the terms $N_{xi}N_y f_i h(f_i N_{xi})$ ($i = 1,2$), which are subtracted from the right-hand sides of Equations (9.10b) and (9.10c) respectively must be considered. This alters the graphical determination of the steady-state values of the autotroph. Now these points, referred to as P_1 and P_2, are the intersections of the g_1/f_1 and g_2/f_2 curves with the $N_y h(N_{x1}, N_{x2})$ curves (straight lines called C_1 and C_2 in Figures 9.4(a,b) and representing consumption). For example, if only autotroph 1 is present, line C_1 starts at the point N_x^* on the axis and rises at a constant slope, $h(f_1 N_x^*)$, as N_y increases.

In Figure 9.4(a) the herbivore feeds proportionally harder on autotroph 1 than on autotroph 2; thus, it pushes $N_{x1} + N_y$ down and N up to levels where autotroph 2 is superior. The point P_1, which is the steady-state equilibrium of Equation (9.10b) is unstable with respect to possible invasion by autotroph 2, because

$$g_2(N_T - N_{x1}^* - N_y^*)/f_2 > g_1(N_T - N_{x1}^* - N_y^*)/f_1.$$

The reverse of this inequality is true at the point P_2, where autotroph 2 is the only autotroph in the system and is at steady state; autotroph 1 can

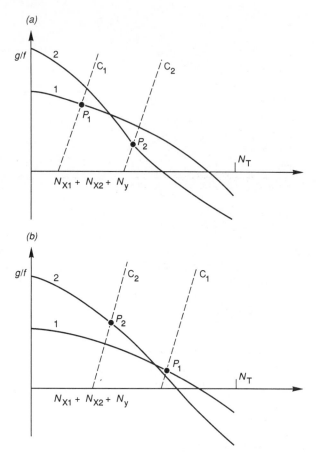

Figure 9.4 Equilibrium points of Equations (9.10a) and (9.10b), P_1 and P_2, when only autotroph 1 and herbivore are present and when only autotroph 2 and herbivore are present. These equilibria are formed by intersections of the g/f curves (1 and 2) and the $N_y h(X_1, X_2)$ curves (C_1 and C_2). In (a) each of these points is invasible by the other autotroph, so there are two alternative stable equilibria. In (b) autotroph 1 cannot invade when autotroph 2 exists at point P_2 and autotroph 2 cannot invade when autotroph 1 exists at point P_1. (Adapted from Armstrong, 1979.)

invade. The fact that both species can invade when the other is present suggests that the species can coexist.

In the second case (Figure 9.4*b*) both P_1 and P_2 are non-invasible steady-state equilibria. This means that these represent alternative stable steady states. Which states exists depends on which autotroph was first to invade.

These mathematical and graphical preliminaries provided Armstrong (1979) with a way of describing successional changes in autotrophs as total nutrient, N_T, in the system is increased. This is illustrated in Figure 9.5(*a*).

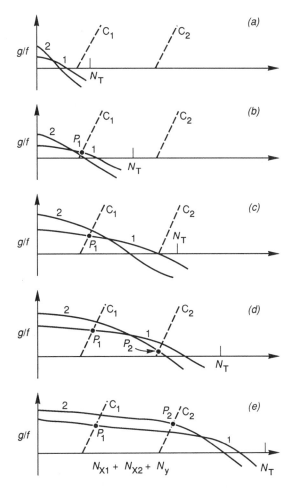

Figure 9.5 The successional changes in a system with two autotrophs (1 and 2), two herbivores, and a limiting nutrient, N. As the total nutrient, N_T, in the system is increased, the configuration of species populations at equilibrium changes (see text for details). (Adapted from Armstrong, 1979.)

In this figure, N_T is very low and neither autotroph 1 nor autotroph 2 can survive. As N_T increases, the curves g_1/f_1 and g_2/f_2 move to the right. Autotroph 1 is the first species that can invade and attain a positive steady state (Figure 9.5*b*) along with the herbivore, which can also invade. Autotroph 2 can invade (Figure 9.5*c*) but could not sustain the herbivore itself. Note that autotroph 2 can only invade in this case when the herbivore is present to limit autotroph 1. In Figure 9.5(*d*), each species can invade in the presence of the other, so that three-species coexistence is probable. Finally, in Figure 9.5(*e*), the nutrient level is high enough that the nutrient-loving autotroph 2 has a growth rate higher than that of autotroph 1 at both points P_1 and P_2, meaning that it will exclude autotroph 1.

This same graphical argument can be applied to a whole community of autotrophs to show the successive replacement of less by more eutrophically adapted species (Armstrong, 1979, for further details).

Changes in autotroph species composition under temporal or spatial gradients in nutrient levels have been observed in many studies. Olsen and Willen (1980) reported on the effects of reduction in phosphorus loading to Lake Vättern, a large Swedish lake, which had been subjected to heavy P discharges for a few decades. At the start of the reduction of P input, phytoplankton species indicative of eutrophication, such as the diatoms *Diatoma elongatum* and *Stephanodiscus hantzschii*, were present, as well as large cryptomonads, green algae such as *Chlamydomonas* and *Scenedesmus* and blue–green algae, especially *Oscillatoria agardhii*. Since the beginning of the study, phytoplankton biomass has decreased in general (Figure 9.6),

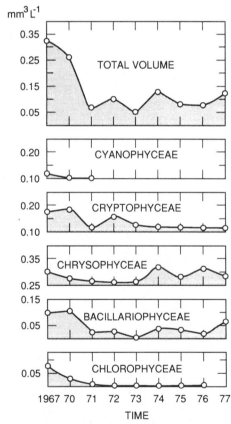

Figure 9.6 Changes in phytoplankton volume in Lake Vättern during the period 1967–1977, based on seasonal means of total volumes of important algal groups. (From Olsen and Willen, 1980.)

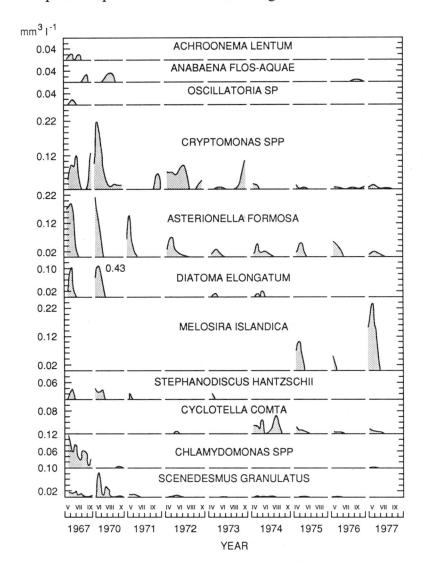

Figure 9.7 Numbers in important phytoplankton species groups in Lake Vättern over the period 1967–1977. (From Olsen and Willen, 1980.)

especially the biomasses of these species (Figure 9.7). Species more indicative of oligotrophic conditions, such as *Melosira islandica* and small crypto-monads, have increased in relative representation.

deNoyelles and O'Brien (1978) enriched experimental ponds with N, P and K to study phytoplankton succession. The *Chrysophyta* group of phytoplankton, initially dominant under oligotrophic conditions, were re-placed by *Chlorophyta* and *Cyanophyta* after only a few weeks. These latter groups of species have higher growth rates than the *Chrysophyta* under high nutrient conditions.

Effects of addition of nutrients to Breckland grass heath were investigated by Davy and Bishop (1984). Nutrient addition stimulated biomass increases in *Festuca ovina* and *Koeleria macrantha*, whereas competing species such as *Hieracium pilosella* decreased.

All of these observations are at least consistent with Armstrong's qualitative conclusions, although they do not prove the model is correct in detail. Armstrong's model assumes that the biotic community can be described by changes in an equilibrium point; that is, the species populations are at equilibrium with respect to each other at any given time, though the equilibrium is continually changing in response to abiotic changes (eutrophication). Equilibrium conditions may, in fact, not necessarily prevail during succession, especially if such processes as eutrophication occur on faster time scales than the response time of the biological populations, and description of succession by a moving equilibrium may not be accurate. However, as a first approximation, this model has great conceptual value.

9.5 MULTIPLE NUTRIENTS

Up to this point, I have been describing the dynamics of systems in which the autotroph is limited only by one nutrient. In actuality, the facts that the same plant may be limited by different nutrients at different times or that neighbouring plants at a given time may be limited by different nutrients may play a fundamental role in the dynamics of plant communities. In fact, an alternative hypothesis to the above mechanism of grazing-induced diversity is that autotrophic growth in general is strongly dependent on the availabilities of at least two limiting nutrients. Such dependence, combined with spatial heterogeneity in each of the limiting nutrients, can explain some of the observed diversity in autotrophs (e.g. Tilman, 1982).

Suppose that two autotrophs, 1 and 2, are limited independently by two nutrient resources of densities N_1 and N_2. Let N_{x1} represent the nutrient equivalent of the biomass density of species 1 and N_{x2} the nutrient equivalent of biomass density of species 2. The equations describing the two autotroph populations are

$$dN_{x1}/dt = N_{x1}F_1(N_1, N_2) - d_1 N_{x1} \tag{9.11a}$$

$$dN_{x2}/dt = N_{x2}F_2(N_1, N_2) - d_2 N_{x2} \tag{9.11b}$$

so that each species has a positive growth rate in the region where

$$F_i(N_1, N_2) - d_i > 0$$

and has a zero or negative growth rate otherwise. If both nutrients are essential, such that the autotroph declines if either falls below some critical concentration, then the $F_i(N_1, N_2) - d_i = 0$ curve for a given species can be represented in the N_1, N_2 plane for one species as shown in Figure 9.8.

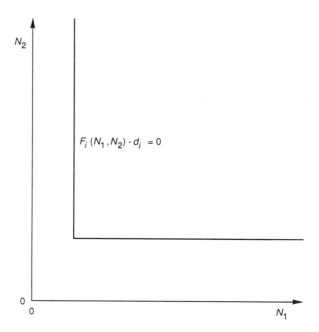

$F_i (N_1, N_2) - d_i = 0$

Figure 9.8 Zero isocline, $F_i(N_1, N_2) = d_i$, for an autotroph, i, limited by two nutrients. When $F_i(N_1, N_2) - d_i > 0$, the autotroph biomass increases. $F_i(N_1, N_2)$ is an increasing function of N_1 and N_2.

When two autotrophs are present in the system, then the outcome of the competition depends, in part, on the relative positions of the two zero isoclines

$$F_1(N_1, N_2) - d_1 = 0$$

and

$$F_2(N_1, N_2) - d_2 = 0$$

Four different cases are possible, as is shown in Figure 9.9:

Case A: autotroph 1 is able to reduce both nutrient concentrations below the levels necessary to sustain autotroph 2 (Figure 9.9a).
Case B: autotroph 2 is able to reduce both nutrient concentrations below the levels necessary to sustain autotroph 1 (Figure 9.9b).
Case C: autotroph 1 is able to reduce nutrient concentration N_1 below the level necessary to sustain autotroph 2 and autotroph 2 is able to reduce nutrient concentration N_2 below the level necessary to sustain autotroph 1 (Figure 9.9c).
Case D: the same as case C but with nutrients N_1 and N_2 reversed (Figure 9.9d).

Cases A and B are trivial in the sense that one species will necessarily

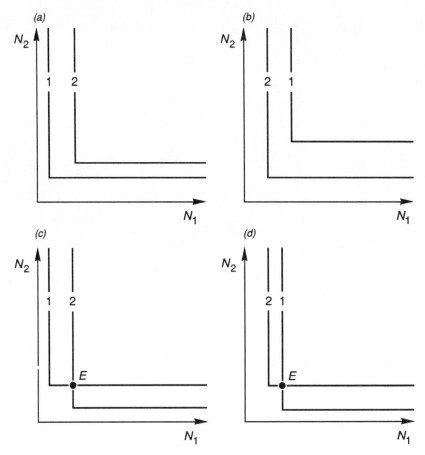

Figure 9.9 Various combinations of zero isoclines of two autotrophs, 1 and 2, limited by two nutrients, N_1 and N_2. See text for discussion. (Adapted from Tilman, 1982.)

displace the other, all other factors being equal. Cases C and D are more complex and interesting. The points of intersection of the two zero isoclines in Figures 9.9(c,d) represent steady-state equilibrium values of N_1^* and N_2^*. The equations for the nutrients are

$$dN_1/dt = I_{n1} - r_1N_1 - N_{x1}F_1(N_1, N_2) - N_{x2}F_2(N_1, N_2) \tag{9.12a}$$

$$dN_2/dt = I_{n2} - r_2N_2 - N_{x2}F_1(N_1, N_2) - N_{x2}F_2(N_1, N_2) \tag{9.12b}$$

If, at the equilibrium point E, either N_{x1} or N_{x2} or both can continue to change, then the equilibrium is unstable. This can be investigated as follows. Suppose, for example, that autotroph 1 has a higher internal ratio $N_1/N_2 = R_1$ (where R_1 is a constant) than does autotroph 2. A slight perturbation that causes an increase in N_{x1} will push N_1 further to the left

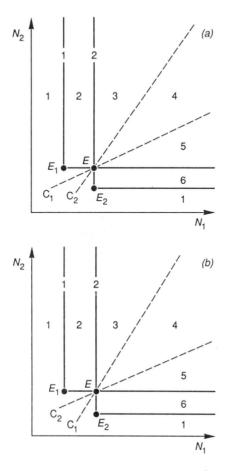

Figure 9.10   The point, E, represents an equilibrium of two autotrophs competing for two limiting nutrients, N_1 and N_2. The dotted lines, C_1 and C_2, represent the ratios at which autotrophs 1 and 2, respectively, take up the two nutrients. In (*a*), the equilibrium, E, is unstable because a deviation from E will amplify through time. In (*b*) E is stable because a perturbation will be resisted by negative feedback. The points E_1 and E_2 are single species equilibria. (Adopted from Tilman, 1982.)

than it will push N_2 down (Figure 9.10*a*). The result will be that the increase in N_{x1} has a greater effect on N_1, the resource that limits autotroph 2, than it does on nutrient N_2, which limits its own growth at equilibrium point E. Hence autotroph 2 will be more affected by negative feedback than will autotroph 1. The resultant decrease in N_{x2} will free up a greater amount of N_2, the limiting resource of autotroph 1 and thus promote its own additional increase. Therefore, for the case in which the autotroph that is limited by a given nutrient N_i at E has a lower ratio of that nutrient in its own biomass than does the other autotroph, a positive feedback cycle will develop, destroying the steady-state equilibrium and causing one or the other of the autotrophs to go to extinction.

Figure 9.10(*b*) portrays the opposite case. Here, when a perturbation causing N_{x1} to increase slightly occurs, N_{x1} depletes its limiting resource, N_2, more than it does N_1, at least compared with what autotroph 2 would do. As a result, the perturbation is resisted by negative feedback and E is stable.

The dotted lines in Figures 9.10(*a,b*) indicate the ratios of nutrients used by autotrophs 1 and 2. When they cross as in Figure 9.10(b), the system will approach the equilibrium point E, whenever the initial nutrient levels are in region 4 of the graph. When the initial nutrient levels are in regions 3 or 5, then the system will approach equilibrium point E_1 or E_2 respectively, representing dominance of autotrophs 1 or 2.

Tilman (1982) concluded from this model that the natural variability of nutrients in soil and water will create conditions in which the species can coexist in some places, whereas one or the other will dominate in other places. It is straightforward to extend this model to any number of competing species. For example, if a third species is added with a zero isocline 3 and a ratio of nutrient uptakes C_3, as shown in Figure 9.11, there is a region 7 in which autotrophs 2 and 3 can coexist.

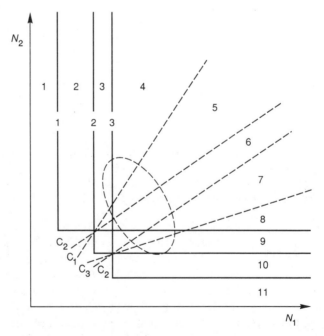

Figure 9.11 Competition of three species for two limiting nutrients. The dashed circle represents the variability of nutrient ratios in a plot of land. All three species can coexist on this plot.

Several experimental studies bear on the predictions of Tilman's model. Tilman and his colleagues (Titman, 1976; Tilman and Kilham, 1976; Tilman, 1977) studied competition between two species of freshwater diatoms (*Asterionella formosa* and *Cyclotella meneghiniana*) under various conditions of silicate and phosphate limitation. The zero isocline of each species was determined, as well as the consumption vectors (Figure 9.12). The model predicts that coexistence should occur when the nutrient levels had initial values in region 4, whereas *Cyclotella* or *Asterionella* should win when those points are in regions 3 and 5 respectively. The results of several experiments are represented in Figure 9.12 as small and large solid circles and large open circles, indicating that the model predicts the outcomes of the competition reasonably well.

Tilman's analysis has important implications for plant diversity. The theoretical implication of models such as Equations (9.1a–c) is that two plant species competing for the same limiting nutrient cannot coexist; the one that can maintain a positive growth rate for the lowest level of the limiting nutrient will displace the others. The rich plant diversity of temperate and tropical forests and of phytoplankton communities in lakes and the ocean cannot be explained by such a model. However, suppose, as can be shown, that there is a great deal of spatial variability in the available concentrations of two critical nutrients in the soil in a plot of land or in a body of water. Then, according to Tilman's model, since each plant type has its own characteristics with respect to these nutrients (e.g. the ambient level

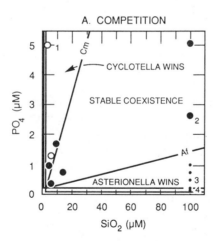

Figure 9.12 Zero isoclines for two diatom species, *Cyclotella* and *Asterionella*, limited by two nutrients, silicate and phospate. The sloping lines represent ratios of the two nutrients consumed by the species. Small solid circles represent starting points for experiments in which *Asterionella* was competitively dominant, large solid circles are for experiments in which the species coexisted, and large open circles are for experiments in which *Cyclotella* was dominant. (From Tilman, 1982.)

at which a nutrient becomes limiting differ among plants; the ratios of various nutrients differ between plants), then there should be spatial regions in which various pairs of plants coexist. Over the whole heterogeneous plot of land a variety of species could coexist in steady state. As a case in point, if, in a plot of land, the variability in ratios of N_1 to N_2 shown enclosed by the dashed circle in Figure 9.11 exist, then the three species 1, 2 and 3 should coexist on that plot.

9.6 SUMMARY AND CONCLUSIONS

The numbers of species in various trophic levels (or the 'widths' of these levels) of a food web are important characteristics of an ecosystem. Despite the fact that each species fills a somewhat unique niche, competition for various resources, including nutrients, generally occurs. Intense competition, which is at least attainable under experimental conditions, can lead to the exclusion of some species by other species that are better in utilizing resources or superior in other ways.

Laboratory and field studies, as well as mathematical models, have been used to determine the conditions under which a number of autotrophs exploiting the same nutrients can coexist. Simple models indicate that the autotroph that is capable of lowering concentrations of a limiting nutrient below that which other autotrophs can use for net positive growth will exclude the others, if all other factors are equal and the system is at equilibrium.

The presence of other factors that differ in effect, such as herbivores that feed differentially on the autotrophs, alters this simple prediction, however, and can allow a whole spectrum of autotrophs to coexist in steady state, even when only one nutrient resource is limiting. Autotroph communities can be changed by changing either levels of available nutrients, the pressure of herbivory, or both.

If the level of a limiting nutrient is gradually increased, either temporally or along a spatial gradient, there will be a sequential replacement of more oligotrophic species (able to utilize low levels of nutrients) by others that can grow faster than their competitors under high nutrient conditions. Such sequences are observed in nature and have been reproduced in the laboratory and with models.

The diversity within the autotroph level in a food web has also been hypothesized to be caused in part by the limitation of autotrophic growth by more than a single nutrient (or by a single limiting nutrient and some other resource). Simple models and experimental studies show that variability of ratios of two limiting nutrients on the fine scale in the environment can allow a great number of autotrophs to be supported at steady state at a given terrestrial or aquatic site.

All of the models described in this chapter are based on the assumption of equilibrium conditions among the biological populations, although in the models the equilibrium point may be slowly changing with time to represent succession, or spatially varying to account for the high levels of autotroph diversity observed in nature. The consistency of the models based on the equilibrium concept with much empirical data at least argues for their plausibility. Still, the applicability of the equilibrium concept to natural systems, which are exposed both to a stochastically varying environment and frequent biotic instabilities (e.g. population cycles and irruptions), has been questioned (DeAngelis and Waterhouse, 1987, for a review). An alternative approach to many ecological phenomena, such as the coexistence of competing species, is based on non-equilibrium concepts. This is discussed in Chapter 10.

10 Temporally varying driving forces and nutrient-limited food webs

10.1 INTRODUCTION

All ecosystems are ultimately the products of exogenous forces of climate and geology, including insolation, temperature and precipitation regimes, wind, overland and groundwater flow and nutrient input. These forces can be termed **driving forces**. They are all at least partly physical in character, but most are strongly influenced by biological structures and processes.

The driving forces govern the rates of a variety of processes that directly build and destroy the physical and biological structures of an ecosystem. If the driving forces were unchanging, an ecosystem might reach a steady-state equilibrium and change very little through time, except for processes of biological evolution. However, the driving forces change continually and, therefore, ecosystems are highly dynamic and always changing and rarely reach a steady state.

The driving forces that shape the food webs of ecosystems vary through time in the following ways:

1. Gradual, continuous changes in external conditions (e.g. geological and climatic changes).
2. Sudden, permanent changes in external conditions. These are often called 'press disturbances', following the terminology of Bender *et al.* (1984) (e.g. change in stream flow regime due to building of a beaver dam).
3. Temporally discrete events or 'pulses' (e.g. fires, storms, droughts, floods). Bender *et al.* (1984) used the term 'pulse disturbances'.
4. Natural periodicities (e.g. annual temperature cycles, annual precipitation cycles).
5. Normal day-to-day variations in physical conditions (e.g. typical fluctuations in windiness and temperature).

Each of these types of temporal variation in the environment can sometimes be a 'disturbance' in the sense of causing direct mortality to organisms (Huston, 1979). However, I will avoid the term disturbance here, as it has too narrow a connotation of something that is harmful to a system. For the most part, these temporal variations are natural and often essential to the ecosystems they affect. The division into five general environmental change categories is somewhat arbitrary and some abiotic forces could fall equally well under one classification or another. For example: (1) a flood can be a disturbance, but in many ecosystems floods are part of the annual hydroperiod and are essential, for example, to the successful breeding of many animals [e.g. the grey teal in Australia (Andrewartha and Birch, 1984)]; (2) whether a change such as a climatic change is gradual or sudden depends on the time scale of interest – the change in mean temperatures and rainfall over the time scale of the lifetimes of individual trees in a forest community would probably be so slight as to be regarded as gradual, but on a time scale over which significant evolutionary changes could occur or on which spatial migration of species could take place, significant climatic change could happen quite rapidly (within a few hundred years) and be regarded as a sudden change.

The view is emerging that temporally varying driving forces play an organizing role as much as they do a disruptive role in an ecosystem (Connell, 1978; Huston, 1979; West et al., 1981; Allen and Starr, 1982; Pickett and White, 1985; O'Neill et al., 1986). A disturbance regime may be instrumental in keeping certain plant and animal types, especially late successional species, from becoming totally dominant. For example, fire slows or prevents the establishment of late successional species that are not fire adapted relative to other faster-growing species. Thus, a tendency towards steady-state equilibrium conditions would imply a drastic reduction in diversity.

The view that ecological communities are usually in a state of disequilibrium changes our perspective on the role of disturbances. Instead of being destabilizing factors, as they are in equilibrium models (e.g. May, 1973), disturbances maintain community diversity. Examples of this point of view are given by Richerson et al. (1970), who postulated that phytoplankton communities are in 'contemporaneous disequilibrium', Levin and Paine (1974), who applied the idea to the marine intertidal, Wright (1974), who discussed the role of fire as a landscape disturbance, and Connell (1978), who interpreted tropical rain forest and coral reef species diversity in terms of disturbance regimes in non-equilibrium systems. The idea can be traced back to Hutchinson (1953, 1961), who proposed that phytoplankton coexistence could be explained in terms of a regime of disturbances that interrupt competitive displacement. Watt (1947) argued even earlier for viewing terrestrial plant communities as mosaics of non-equilibrium patches, each of these changing through successional stages and being reset to earlier stages by disturbances.

Huston (1979) gave these concepts quantitative form in a mathematical model. The basic assumption of this model is that, in the absence of disturbances, competitive displacement, described in the model through Lotka–Volterra competitive interaction terms, acts through successional time to eliminate all but the dominant competitors, which have high carrying capacities, implying efficient use of resources at low levels. The competitively inferior species (low carrying capacities), however, have faster growth rates in the model. Disturbances periodically reduce all of the competitors by equal amounts. The results of simulations of the model showed that the outcomes depended on the frequency of disturbances relative to the rates at which the dominant species competitively eliminated other species. Too low a rate of disturbances, or very high growth rates, resulted in only a few dominant species remaining, while too high a rate of disturbance or very low growth rates prevented the slow-growing dominants from becoming established at all. Thus, diversity was controlled not by a competitive equilibrium between balanced competitive abilities, but by a 'dynamic equilibrium' between competitive exclusion and disturbance. Models similar to Huston's in which temporal variability in survival permitted prolonged species coexistence were a model of a marine benthic community (Woodin and Yorke, 1974) and a model of lottery competition between coral reef fish (Chesson and Warner, 1981).

In this chapter environmental changes and disturbances involving nutrient interactions with food webs are surveyed. These changes may involve both permanent and pulse type changes to nutrient fluxes as well as disturbances that kill organisms and affect nutrient fluxes.

10.2   TEMPORAL SCALES AND RESPONSE TO ENVIRONMENTAL CHANGE

To understand the types of responses to a fluctuating environment, such as fluctuations in nutrient input, it will be useful to consider a simple example of a system driven by a nutrient input (Figure 10.1). The model here has only one compartment, which can be thought of as the aggregated nutrient stored in the system. Two different cases of model input are considered.

In case 1 a square pulse of nutrient input is added on top of the normal input, I_0 (Figure 10.2a). Nutrient is assumed lost from the system at a rate proportional to the stored nutrient $N(t)$. The equation representing the system is

$$dN/dt = -kN + I_0 + \text{SQUARE}[\varepsilon, \alpha I_0] \qquad (10.1)$$

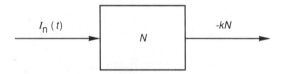

Figure 10.1   A one-compartment system representing stored nutrient, N, driven by a time-varying input, $I_n(t)$.

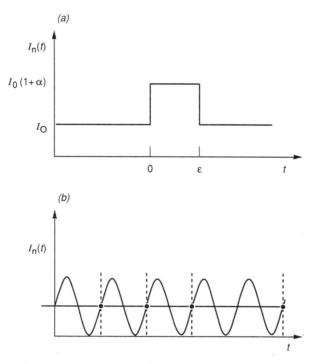

Figure 10.2 Time-varying nutrient inputs to the system pictured in Figure 10.1 (*a*) square wave perturbation and (*b*) sinusoidal wave.

where SQUARE[$\varepsilon, \alpha I_0$] represents a square pulse of amplitude αI_0 (where α is assumed to be the order of magnitude of unity) and duration ε.

The solution of the model equation for times t larger than ε and for an initial value equal to the steady-state value, $N(0) = I_0/k$, is

$$N(t) = I_0/k + (\alpha I_0/k)\,\exp(-kt)[\exp(k\varepsilon) - 1] \tag{10.2}$$

The response of the stored nutrient, $N(t)$, to the pulse depends on the loss rate coefficient, k, which is also the eigenvalue of Equation (10.1). If k is small enough that $k\varepsilon \ll 1$, then the maximum value reached by $N(t)$, which occurs at $t = \varepsilon$, is approximately

$$N(\varepsilon) \cong (I_0/k)(1 + \alpha k\varepsilon) \tag{10.3}$$

which represents a deviation of $(I_0/k)(\alpha k\varepsilon)$ from the steady state, which is very small. If k is large enough that $k\varepsilon \gg 1$, then the maximum value of $N(t)$ again occurs at $t = \varepsilon_j$

$$N(\varepsilon) = (I_0/k)(1 + \alpha) \tag{10.4}$$

In the first case, $k\varepsilon \ll 1$, the variable N 'feels' virtually no effect of the nutrient pulse, whereas, in the second case, where k is large enough that

$k\varepsilon \gg 1$, N fully responds to the pulse. For intermediate values of k, the response of N lies somewhere in between.

If the duration ε of the pulse is extended for long enough (by turning the pulse into a permanent change in resource level), then no matter what size k is, eventually the steady state

$$N^* = (I_0/k)(1 + \alpha) \tag{10.5}$$

is reached. The smaller k is, the longer it will take to reach this new steady state.

The above results have some rather general implications despite their basis on a quite simple model. The parameter k, which in mathematical terms is the eigenvalue of Equation (10.1), is related inversely to the 'resistance' of the system to this type of perturbation; the smaller k is, which is to say, the larger is the characteristic decay time ($T_R = 1/k$) the more resistance the system has to short-term changes in the input I_0. The actual mechanism for resistance in this example is straightforward. A small value of k means that the steady-state nutrient amount in the unperturbed system, $N^* = I_0/k$, is large. Thus, any perturbation in the form of a pulse of elevated or decreased nutrient input must be continued over a long duration to have a significant relative effect on N^*. In more general systems, however, resistance can result from any number of mechanisms. Some of these will be mentioned later.

In addition to the simple discrete pulse input studied above, the input can also be varied to simulate some natural periodicity, say a sinusoidal function with period $\tau = 2\pi/\omega$ of amplitude I_0 about the mean input I_0 (Figure 10.2b). The equation representing this system is

$$dN/dt = -kN + I_0[1 + \sin(\omega t)]. \tag{10.6}$$

The constants $T_R = 1/k$ and τ define two time scales of the system.

The exact solution of Equation (10.6), when $N(t)$ has an initial value of N_0, is

$$N(t) = I_0/k + [I_0/(k^2 + \omega^2)][k \sin(\omega t) - \omega - \cos(\omega t)]$$
$$+ N_0 \exp(-kt) \tag{10.7}$$

Two interesting extreme cases are the following:

Case 1. $\tau \gg T_R$ (or $k \gg \omega$). This represents a very gradual, continuous change in the nutrient input. In this case, for $t \gg T_R$

$$N(t) \cong I_0/k + (I_0/k) \sin(\omega t) \tag{10.8}$$

so the stored nutrient tracks the input (Figure 10.3a). The system in this case 'feels' the changing input and is always in equilibrium with it.

Case 2. $\tau \ll T_R$ (or $k \ll \omega$). This case represents a temporally varying nutrient input that oscillates so rapidly that the system does not respond to

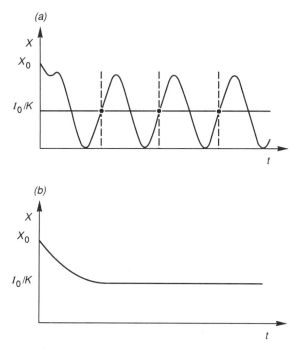

Figure 10.3   Response of N in the system of Figure 10.1 to a sinusoidal input when (*a*) $\tau \gg T_R$ and (*b*) $T_R \gg \tau$.

the alternations in magnitude, in the same sense that ocean temperature does not respond appreciably to the diurnal temperature cycle. As far as the dynamics of the system are concerned, the fluctuations may as well not be occurring. The second term in Equation (10.7) scales as $1/\omega$ and is thus so small compared with I_0/k, that the oscillations in X are negligible in magnitude. The stored nutrient variable merely moves from its initial value to the equilibrium value, I_0/k (Figure 10.3*b*).

Case 3. The intermediate case, when $\tau \cong T_R$, is the most interesting, because the system is influenced by the changing input but never comes to equilibrium with it. This type of situation has implications for the coexistence of competitors in multispecies communities, because it indicates that temporally varying external driving forces can prevent the system from approaching equilibrium, and thus in some cases competitive exclusion, from occurring. This will be examined in more detail later.

The basic concepts here have been derived for a simple linear one-compartment system. However, they provide a good starting point for studying the response of more complex systems to environmental variations. In particular, they illustrate the concept of scale. All variations have some temporal scale (e.g. the duration ε of a pulse or the period τ of an oscillation) and the relationship of this temporal scale to the characteristic temporal scale of the system determines the effect.

10.3 RESPONSE OF FOOD CHAIN MODELS TO PERMANENT CHANGES IN NUTRIENT FLOWS

The response of whole food chains or ecosystems to environmental changes, such as pulse or press disturbances to the nutrient input, has been the object of considerable interest and modelling analysis (e.g. Child and Shugart, 1972; Webster *et al.*, 1975; Harwell *et al.*, 1977; Jordan *et al.*, 1972; Harrison, 1979; Harrison and Fekete, 1980). There has been particular emphasis on comparing ecosystems of different types in their resistance to externally imposed changes or fluctuations.

The concept of resistance of a system to disturbances is complex. As Harrison and Fekete (1980) point out, resistance can be interpreted as the response (or, more precisely, the ability to resist change) following the imposition of some form of stress, such as harvesting, drought, pollutant input and so forth. One such stress would be permanent alteration of one or more of the nutrient flows in an ecosystem.

Harrison and Fekete (1980) worked out the mathematical details for calculating the dynamics of change in a linear ecosystem model following the imposition of any combination of changes in flow parameter values. This mathematical approach, while complex, is actually only an elaboration of the analysis of the one-compartment system in the preceding section. The eigenvalues of the model again play a critical role in determining how the system will behave. For an n-compartment model, there are n eigenvalues, so the possible modes of behaviour are more complex and can possibly span a wide range of time scales, as opposed to the single characteristic time scale of the one-compartment model. This will be illustrated by specific examples.

Harrison and Fekete (1980) applied their analysis to eight linear models of nutrient flow in ecosystems, which were originally constructed by Webster *et al.* (1975) (Figure 10.4). These models include a variety of ecosystems, but all have the same basic set of components. The analysis applied by Harrison and Fekete included determination of the eigenvalues of each system, three of which for each system are shown in Table 10.1. All of the eigenvalues (in units of year^{-1}) are real (except in the case of the lake model, where there is a pair of complex eigenvalues) and negative, meaning that the system is stable about the steady-state equilibrium. The critical root, λ_1, is the one with the negative value closest to zero (that is, the one with smallest absolute value), so it is the slowest characteristic time associated with the system. As an example of how one can interpret these eigenvalues, consider the tropical forest, which has eigenvalues over the range from -49.90 year^{-1} to -0.000420 year^{-1}. This means that some combination of com-

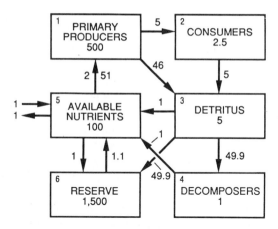

Figure 10.4 Flow diagrams for the hypothetical nutrient cycling system of one (the tundra ecosystem) of the eight different types of ecosystems constructed by Webster *et al.* (1975). There are six compartments: (1) primary producers, (2) consumers, (3) detritus, (4) decomposers, (5) available nutrients, (6) reserve. (From Harrison and Fekete, 1980.)

Table 10.1   Three of the eigenvalues for each of the ecosystem models shown in Figure 10.4. Eigenvalues λ_1 and λ_2 are the two eigenvalues smallest in absolute value, λ_1 being the critical eigenvalue that governs the long-term response of the system. Eigenvalue λ_6 is the eigenvalue with the largest absolute value (from Harrison and Fekete, 1980)

System	λ_1	λ_2	λ_6
1. Tundra	−0.00152	−0.0121	−2.50
2. Grassland	−0.000358	−0.0160	−5.505
3. Temperate forest	−0.0000622	−0.00668	−5.50
4. Tropical forest	−0.000420	−0.00375	−49.90
5. Ocean	−0.762	−1.57	−105.5
6. Lake	−0.00831	−0.709	−20.4 + 1.03i*
7. Salt marsh	−0.00135	−0.316	−19.75
8. Stream	−0.0999	−2.00	−1010.0

*Note that eigenvalue λ_6 for the lake system has an imaginary part $+1.03i$, meaning that there is some oscillatory tendency (though it is strongly damped by the real part, -20.4). Whenever there is a complex eigenvalue, such as λ_6, there is also the complex conjugate, which is $\lambda_7 = -20.4 - 1.03i$.

ponents of the tropical forest ecosystem model will approach steady-state equilibrium on a time scale of 1/49.90 (or about 0.02) years, while other parts of this ecosystem model will take 1/0.00042 (or about 2400) years. Sometimes a particular eigenvalue may be associated almost completely with one compartment. In this case, for example, a very slow soil nutrient reserve compartment may primarily account for the smallest eigenvalue, while the consumer component, having rapid nutrient turnover, may determine the eigenvalue that is largest in absolute value. In general, however, since the eigenvalues are products of the whole system, it is impossible to associate specific eigenvalues precisely with specific compartments.

To compare the behaviour of the eight systems, Harrison and Fekete (1980) performed the following changes on each model system. They (1) permanently increased by 5% the outflow, F_{05}, from compartment 5 (available nutrients), which had been normalized to have the same initial value of $N_5 = 100$ for all systems to facilitate comparison. This 5% change in parameter value altered the eigenvalues slightly, but not enough to need to be taken into consideration in the computations. The authors used their eigenvalue analysis to calculate the changes in compartment 5 through time following the imposition of the change. Figure 10.5 shows the changes in $N_5(t)$ for each ecosystem type over a 3000-year period.

What is most striking in Figure 10.5 is the wide variation in time scales of responses. The stream ecosystem reached a final changed value of $\Delta N_5 = -4.76$ (corresponding to a 4.76% decrease from the original steady-state value of $N_5 = 100$) within a few years, the lake ecosystem required over 200 years to approach the same value, and the grassland, tropical forest, temperate forest and tundra systems required thousands of years. Note that the early change in the temperate forest (determined by the larger eigenvalues) was faster than that of the tundra, tropical forest and grassland, but that its long-term change (dictated by the smallest or critical eigenvalue) was slower than the others. It is not too surprising that several of the systems

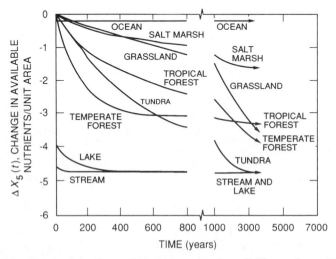

Figure 10.5 Graph of the change in compartment 5, available nutrients, N_5, versus time after the outflow from compartment 5 is increased by 5%, for the model systems of Figure 10.4. (From Harrison and Fekete, 1980.)

approached the same final deviation of -4.76, because this final value is determined by the change in the rate of outflow from compartment 5. The ocean and salt marsh ecosystems reached final deviations different from -4.76. This was the result of the fact that these systems have other outflows besides that from compartment 5 (the stream system also has additional outflows, but these are very small compared with that from compartment 5, so the change in that outflow dictated that ΔN_5 for the stream also approached a deviation of approximately -4.76).

From our discussions of resilience and turnover time in Chapter 2 and subsequent chapters, one would expect that the rate at which the long-term change in each of these ecosystems is approached will vary in proportion to the nutrient turnover or residence time, T_{res}, which is the ratio of the outflow of nutrient from the system to the total nutrient stored in the system in steady state. Thus, for the tropical forest, this ratio is $1/(500 + 2.5 + 5 + 1 + 1500 + 100) = 1/2108.5 = 0.00047$, which is close to the absolute value of the critical eigenvalue for the tropical forest, 0.00042 (Table 10.1).

Resistance in this example is seen to be characterized not only by the absolute magnitude of change, but also by the time period considered. The ocean model is most resistant in the sense of the small relative magnitude of its ultimate change resulting from a change in the flow out of compartment 5; however, it is by far the least resistant in terms of the rapidity of its reaching this final changed state. Note that its smallest eigenvalue, λ_1 is -0.762, much larger in absolute value than the others. The temperate forest would be judged one of the least resistant systems when looked at on a time scale of 200 years, but it is more resistant than most on a scale of 3000 years.

Resistance is a complex concept from another point of view also. Harrison and Fekete (1980) measured resistance by the relative change in compartment 5, the available limiting nutrient, following a permanent change in the flux out of compartment 5. However, many other fluxes could be altered, and changes in other compartment sizes could just as logically be chosen as measures of resistance. In fact, the authors also changed the flux from compartment 5 to compartment 1, F_{15}, instead of the outflow, F_{05}, and looked at changes in total system nutrients that resulted. The results from changing F_{15} are much more complex. Rather than changing monotonically through time, the changes can go through different phases as different eigenvalues pass in and out of control over the dynamics through time (Figure 10.6). Thus, the particular parameters and variables chosen can make an enormous difference to what value of resistance is ascribed to the system.

10.4 TEMPORAL VARIABILITY AND THE PERSISTENCE OF COMPETING SPECIES

A classic result of mathematical modelling of ecological communities is that two or more species cannot coexist stably on a single resource. This result

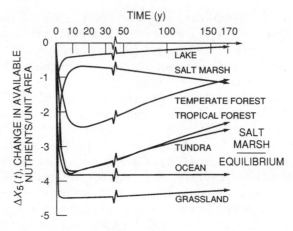

Figure 10.6 Graph of the change in compartment 5, available nutrients, ΔN_5, versus time after the flow from compartment 5 to compartment 1 is increased by 5%, for the model systems of Figure 10.4. (From Harrison and Fekete, 1980.)

was discussed in Chapter 9, but to briefly review it, consider two autotrophs, 1 and 2, characterized by equivalent nutrient amounts N_{x1} and N_{x2}, competing for a limiting nutrient, N, as described by the equations:

$$dN/dt = I_n - r_n N - r_1 N N_{x1} - r_2 N N_{x2} \qquad (10.9a)$$

$$dN_{x1}/dt = r_1 N N_{x1} - d_1 N_{x1} \qquad (10.9b)$$

$$dN_{x2}/dt = r_2 N N_{x2} - d_2 N_{x2} \qquad (10.9c)$$

Simplified Lotka–Volterra interactions between the autotrophs and the nutrient are used here for convenience of the analysis. The parameters are all assumed to be constant for the moment.

The Equations (10.9b,c) imply individually that the equilibrium value of N should be

$$N^* = d_1/r_1 \qquad (10.10a)$$

and

$$N^* = d_2/r_2 \qquad (10.10b)$$

which, because d_1/r_1 is not generally equal to d_2/r_2, are inconsistent; N cannot be both values at once. In fact, if

$$d_1/r_1 < d_2/r_2 \qquad (10.11)$$

then N_{x2} will go to zero in the model, a mathematical expression of the extinction of one of the autotrophs from the system. The reason that competitor 2 goes to extinction is very simple. Competitor 1 holds the available nutrient at a level below the threshold for positive growth of competitor 2. The latter species necessarily declines to zero. If the inequality (10.11) were reversed, then N_{x1} will go to zero. This mathematical reasoning can be extended to more general systems of n consumers to show that they cannot coexist on $m < n$ resources.

This basic result in its earliest form was due to Volterra (1931) and is referred to as competitive exclusion. This mathematical result has engendered much discussion and controversy in ecology. Some ecologists hold that it is an important principle that forbids species from being specialized on the same resource, while others argue that this is not really a theory but a tautology. Whether it is a tautology or not, however, the result has great operational value and has stimulated a search for mechanisms that can explain coexistence of a large number of autotroph species on a few nutrient species. In Chapter 9, both the effects of higher trophic levels and the effects of spatial heterogeneity of two limiting nutrients in maintaining species diversity were considered.

This result has been reexamined from another perspective by Armstrong and McGehee (1980) among others. A key conclusion of these theoretical studies is that mathematical competitive exclusion does not necessarily hold when the variables are not constant at an equilibrium but are temporally varying, an observation made by Hutchinson in 1953.

To discuss the effects of temporal variability on ecological communities, it is useful first to introduce the concept of persistence. 'Persistence is the tendency for the components of a system to stay within specific bounds through time, although these components may never approach constant values. For example, a nutrient-limited autotroph–herbivore food chain may oscillate (that is, exhibit local instability), while the components stay above certain population or biomass levels' (DeAngelis et al., 1989). The importance of persistence on ecosystems, as opposed to local stability (absence of oscillations) was noted by Botkin and Sobel (1975), Hallam (1978) and others.

Can temporal variations in one or more of the parameters of Equations (10.9a,b,c) promote the coexistence of the two competing autotroph species 1 and 2? Levins (1979) studied this problem and showed through some simple mathematical arguments that temporal variability alone could lead to coexistence in the case where one species would tend to exclude the other under constant conditions, but only when the system had certain types of non-linearities.

As an example, Levins (1979) replaced the equation for the second autotroph, Equation (10.9c), by

$$dN_{x2}/dt = [r_2(1 + \beta N)N - d_2]N_{x2} \tag{10.12}$$

and made corresponding changes in Equation (10.9a), so that it became

$$dN/dt = I_n - r_n N - r_1 N N_{x1} - r_2(1 + \beta N)N N_{x2} \tag{10.13}$$

He assumed that

$$d_2/r_2 > (1 + \beta d_1/r_1)d_1/r_1 \tag{10.14}$$

so that competitor 2 is competitively excluded under conditions in which the parameters are all constant. Autotroph 1 holds the nutrient level too low for the right-hand side of (10.12) to reach equality, so $dN_{x2}/dt < 0$. Suppose, however, that the resource level fluctuates randomly; then the equilibrium condition for Equation (10.12) is found by time averaging over the right-hand side of (10.12) and setting it equal to zero:

$$r_2(1 + \beta \bar{N})\bar{N} + r_2\alpha \, \mathrm{var}\,(N) = d_2 \tag{10.15}$$

where $\bar{N}$ is the mean of N and var (N) is its variance. The presence of the variance term in addition to the mean value $\bar{N}$ emerges from the time average over $r_2\beta N^2$, where N is a random variable. The existence of the variance term in (10.15) means that there is the possibility, at least, of Equation (10.15) being satisfied independently, at the same time that Equation (10.9b) is satisfied. This depends on a sufficiently large value of var (N) occurring. As was shown by Levins, some temporal variation in nutrient input, I_n, can cause a positively correlated variance in N. This variance is a variable that under some circumstances can satisfy Equation (10.15).

An intuitive explanation of this phenomenon is the following. The non-linear term $\beta r_2 N^2$ means that competitor 2 is more effective than competitor 1 at feeding on high levels of resource N. In steady-state conditions in which N is constant, competitor 1 would depress N to a value d_1/r_1 that is low enough to cause competitor 2 to go to extinction. When the nutrient input parameter I_n or other parameters vary temporally, however, N will fluctuate to both lower and higher values, and at the latter competitor 2 has an advantage over competitor 1. The variance in these fluctuations is itself variable and depends on the mean level of N_{x2}, since competitor 2 will tend to 'trim down' the high fluctuations of N. As long as inequality (10.14) is satisfied (meaning that species 1 displaces species 2 in the absence of temporal variations), and as long as d_2/r_2 is not too much larger than $(d_1/r_1)(1 + \alpha d_1/r_1)$ so that values of var (N) can feasibly be of the right order of magnitude to satisfy Equation (10.14), then a temporally fluctuating environment will maintain both competing autotrophs. To demonstrate this, a computer simulation was performed on a system similar to Levins' (DeAngelis et al., 1985). The parameter values are such that without the fluctuations in I_n, N_{X2} would go rapidly to zero, as shown in Figure 10.7(a). However, when random fluctuations are applied to N, species 2 tends to persist (Figure 10.7b).

Levins' (1979) results have been discovered independently by others. Kemp and Mitsch (1979) were motivated by the 'paradox of plankton' noted earlier by Hutchinson (1953, 1961). Hutchinson pointed out that a typical lake epilimnion, which may be assumed to be reasonably homogeneous, has

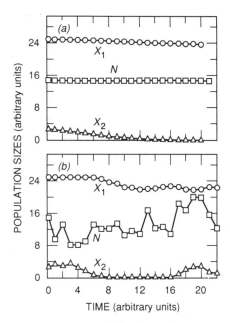

Figure 10.7 Computer simulations of Equations (10.9a,b) and (10.12) of Levin's model of two consumers, N_{X1} and N_{X2}, feeding on a single resource, N. (a) No fluctuations in N; (b) the resource fluctuates randomly and uniformly over the interval from $0.5N$ to $1.5N$, where N is the average value of nutrient. (From DeAngelis *et al.*, 1985.)

many different coexisting phytoplankton species, most of which must be in competition for a small set of nutrients, light and other resources. According to the arguments of competitive exclusion discussed earlier, most of these species should go to extinction. Hutchinson reasoned that this did not occur because environmental disturbances acted frequently enough to stop a competitively superior species from overgrowing the others. Kemp and Mitsch tested this, though not, like Levins, through mathematical arguments, but through a computer simulation model of three competing phytoplankton species populations. The model is fairly realistic, especially in its description of the effects on phytoplankton of hydrodynamic variables. Effects of water movement can enhance nutrient uptake as well as create stresses on the cells. Phosphorus was modelled explicitly as the limiting nutrient. Functions for solar radiation, self-shading and nutrient regeneration in detritus were included (Figure 10.8). The model system was subjected to water movement having different levels of kinetic energy and having different frequencies from the spectrum of turbulence (representing different-sized eddies) over a range from $30\,d^{-1}$ to $0.01\,d^{-1}$.

Under quiescent conditions, only one species (Species 3) persisted. When kinetic energy was applied at the high and low frequency extremes ($0.01\,d^{-1}$

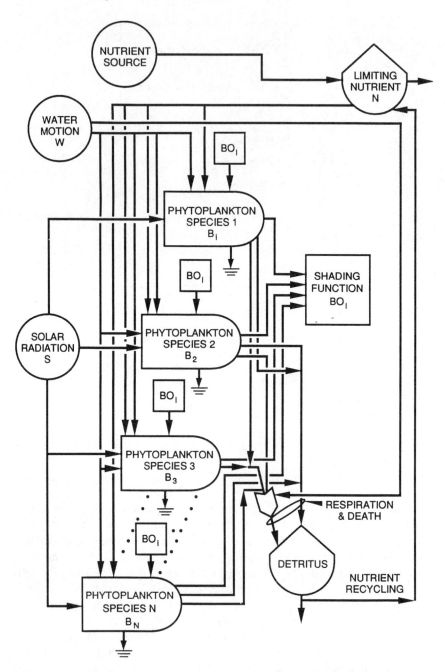

Figure 10.8 Diagram of model to investigate relationships between the outcome of competition between three phytoplankton populations and temporal variability through wave motion. (From Kemp and Mitsch, 1979.)

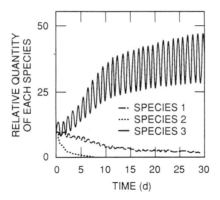

Figure 10.9 Simulated behaviour of three phytoplankton populations under conditions of a turbulent hydrodynamic regime with a primary frequency of $0.01\,\mathrm{d}^{-1}$.

and $30\,\mathrm{d}^{-1}$ respectively), some weak coexistence of Species 1 was achieved along with Species 3 (see Figure 10.9 for a typical simulation for the lower extreme). For reasons discussed in the second section of this chapter, the high and low frequency oscillations did not affect the community significantly. The populations did not respond to the high frequency perturbations. The low frequency perturbations affected the populations, but not enough to keep competitive exclusion from occurring. Finally, an intermediate frequency of $1.0\,\mathrm{d}^{-1}$ maintained all three species (Figure 10.10). This frequency approached the turnover rate of plankton, so, as discussed earlier, this frequency can keep populations away from equilibrium and prevent the competitive exclusion that could take place under equilibrium conditions.

These results relate to those of Levins (1979). They indicate that in order

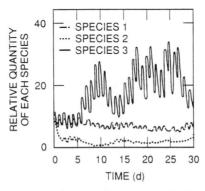

Figure 10.10 Simulated behaviour of three phytoplankton populations under conditions of a turbulent hydrodynamic regime with a primary frequency of $1.0\,\mathrm{d}^{-1}$. (From Kemp and Mitsch, 1979.)

for a variance in the resource, var(N), to occur that is able to facilitate coexistence of competitors 1 and 2 in Levins' model, it is necessary that the driving force that causes variations in I_n or one of the other parameters be in the appropriate frequency range.

The model examples of both Levins (1979) and Kemp and Mitsch (1979) involve the application of external temporal variability. Another possibility is that fluctuations may be internally generated through limit cycle oscillations of competing species that can occur even when the driving forces are all constants. Armstrong (1976) showed that several competing species could coexist on a smaller number of abiotic resources when the system tends to stable periodic orbits.

Is there observational or experimental evidence for these theoretical results? Experiments by Sommer (1984) seem to confirm that a temporally varying nutrient input can maintain phytoplankton diversity. Sommer inoculated 20-litre chemostats with a natural assemblage of phytoplankton. The nutrient solution contained no phosphorus. Phosphate was added to this solution in weekly pulses. Under the oscillatory conditions, a more diverse array of phytoplankton was able to persist than under steady-state controls in which the phosphorus input was constant. Suttle *et al.* (1988) subjected a freshwater phytoplankton assemblage to nutrient addition pulses on time periods raging from 4 to 16 days. Different periodicities led to differences in communities and in average sizes of cells. One explanation advanced for the effects of nutrient pulsing on communities was that individual species are specialized for using different levels of a fluctuating resource. Some species, for example, may be more effective than others at utilizing temporary high peaks in nutrient availability.

Variation in nutrient supply can also influence greater species diversity at higher trophic levels. Neill (1984) showed that periodic nutrient pulses increased diversity of herbivore trophic levels. In a small oligotrophic mountain lake, cladoceran species normally outcompeted rotifers. When the lake was subjected to large biweekly pulses of nutrient, however, high rotifer populations could be maintained in coexistence with the cladocerans.

Walsh (1975) suggested that different turbulent spectra in upwelling and oceanic areas could lead to different food chain structures because of their effects on feeding and metabolic processes of herbivorous consumers. Walsh found from spectral analysis of turbulence that a predominant period in coastal waters is about 2 to 10 days, whereas in oceanic waters it is approximately 0.5 to 2 days. For coexistence to occur, the frequency of fluctuations must be matched to the rate at which population growth and competition lead to competitive exclusion. As a system is enriched and population growth rates increase, the frequency of environmental variation must also be increased in order to allow coexistence. This is analogous to the relationship between the rate of competitive exclusion and disturbance frequency in Huston's (1979) dynamic equilibrium model.

10.5 EFFECTS OF LARGE ENVIRONMENTAL CHANGES

A question of importance is whether a sufficiently large perturbation can completely change the state of a system over a long period of time. The possibility of such a switch in system type exists only in systems that are highly non-linear. A linear model will always return to the same equilibrium point, assuming this point is stable, following a perturbation of any size. In a non-linear system, however, there is a possibility that other stable equilibria with different species compositions exist (e.g. Holling, 1973). A large perturbation may push a system away from the domain of attraction of one stable equilibrium and into the domain of attraction of another alternative equilibrium state. This can occur when the perturbation is large enough to trigger positive feedbacks that cause the deviation from the original state to continue to grow.

Clearcutting of a forest is an example of a major disturbance having a variety of secondary effects that can result in system-level change. Clearcutting normally leads to increased losses of nitrates and other nutrients due to increased runoff of water (Vitousek and Melillo, 1979; Bormann and Likens, 1979). Nitrogen loss is especially likely because of the large amount of nitrogen that is recycled through the soil and back to the plants per unit time relative to the standing pools of available nitrogen (nitrate and ammonium in the soil). Forest harvesting both interrupts the uptake of inorganic nutrients (which are most susceptible to losses through leaching) in plants and results in faster rates of decomposition and nitrogen mineralization (Boring et al., 1981; Matson and Vitousek, 1981; Gordon and Van Cleve, 1983). The inorganic nutrient pools may then be exposed to leaching for a number of years.

These conditions would appear appropriate for a complete change in the ecological system, as a sufficiently large loss of nutrients could prevent the reestablishment of the original forest community. In fact, only under certain circumstances does a total system change occur, because the forest soil system has properties that tend to resist the continued losses of nutrients. In an examination of 17 forest sites in North America, Vitousek (1982) concluded that nitrate losses from disturbed sites were controlled by three main factors: '(1) the amount of nitrogen mineralized annually prior to disturbance, and the extent to which that mineralization is increased by disturbance; (2) the interaction of soil processes, particularly the immobilization of nitrogen in organic form and delayed nitrate production, which can keep mineralized nitrogen in relatively immobile forms, and (3) vegetation growth and the reestablishment of plant nitrogen uptake in disturbed sites'. On sites where nitrogen was limiting prior to the disturbance, carbon/nitrogen ratios in plant remains are high and the microorganisms of decomposition are effective at immobilizing large amounts of nitrogen (Bosatta and Staaf, 1982; Chapter 7). These microorganisms can also outcompete autotrophic nitrifiers for ammonium and thus delay nitrate production. This immobilization and delay in nitrification of ammonium to nitrate can substantially control the losses of nutrients, as long as a sufficient amount of plant residue is left on the site to sustain the microorganisms that immobilize the nutrients.

In environments that are close to the margins of a particular community type, however, the effect of clearcutting may be to produce irreversible changes. Perry *et al.* (1989) have hypothesized that the difficulty in restoring clearcut fir forests at high elevations on granitic soils in northwestern North America may be an example of this. The cutting down of plants reduces the input of energy and nutrients from root exudates. More of the ammonium pool is hypothesized to be oxidized by autotrophic bacteria into nitrate, which is susceptible to denitrification and leaching. Nitrogen immobilized by bacteria may be less available to fungi that are mutualistic with the original fir trees because of the reduced input of energy from root exudates. These changes promote a number of other changes in the soil system, including changes from fungal to bacteria dominance and an increase in soil pH, which feed back positively in reinforcing the effects of the perturbation. Perry and his colleagues suggest that in this type of ecosystem, as well as in others, such as the Miombo woodland of southern Africa, the impact of heavy harvesting or clearcutting breaks vital links of nutrient recycling and leads to degradation of the original vegetation to annual grasses.

Possible cases in which a large perturbation in nutrient levels could result in a drastic change of state have been proposed by Stachurski and Zimka (1975), Gosz (1981), Vitousek (1982) and Pastor *et al.* (1984) for ecosystems with woody vegetation. A disturbance resulting in a strongly decreased nitrogen availability in the soil might cause some of the woody plants to increase the efficiency of their use of nitrogen, including the production of litter with low levels of nitrogen relative to carbon. Mineralization of nutrients from such litter is extremely slow, because nitrogen, the limiting nutrient resource, is initially immobilized by decomposers. This could exacerbate the nitrogen deficiency in the soil and cause the plants to utilize nitrogen even more efficiently and produce even lower quality litter. As Vitousek pointed out, both phenotypic and genotypic changes in the vegetation may occur. The result can be a self-reinforcing (positive feedback) change towards worsening nutrient conditions and a new vegetation community adapted to these conditions.

Some empirical support exists for these theoretical conclusions. A specific example of two divergent types of community that are likely to exhibit changes of state when subjected to nutrient perturbations are (1) the white spruce and hardwood type and (2) the black spruce type. These communities, both of which can occupy sites in the taiga zone, are characterized by different nutrient cycling efficiencies (Van Cleve *et al.*, 1983). Black spruce ecosystems have mechanisms that conserve nutrients. Because the litter of black spruces is low in nutrients, it is slow to decompose; thus, thick layers of organic matter accumulate on the forest floor under black spruces. This insulates the soil and thus reduces summer soil temperature. Rates of mineralization become even slower on these stands because of low summer temperatures. It is possible that when a white spruce and hardwood system suffers a sharp reduction in nutrient availability, the deviation will be reinforced as black spruces invade and further reduce nutrient availability. A positive feedback between black spruce vegetation type and low nutrient availability may precipitate a shift in species composition and nutrient dynamics. A change in ecosystem structure and dynamics could be precipitated.

In aquatic systems, pulses of nutrients can cause major changes in ecological communities. In one case, already discussed in Chapter 8, Neill (1988) experimentally explored the stability of food webs in oligotrophic lakes. In another example, Hessen and Nilssen (1986) performed enclosure experiments in a eutrophic lake. Perturbations were performed by additions of nutrients, fish, herbivorous zooplankton, or combinations of these. Additions of nutrient increased both algae and bacteria, the latter probably because of the increase in algal exudates. Addition of fish plus nutrients caused a shift towards smaller, less efficient grazers, and higher levels of algae and bacteria. Figure 10.11 represents the observed types of shift for different perturbations.

Like lakes, streams can be affected easily by changes elsewhere in the watershed, such as clearcutting of surrounding forest. Dependence on allochthonous carbon and other nutrient inputs is much stronger in streams than in lakes, however, since the proportion of new to recycled nutrients is much higher in streams. Webster et al. (1975) suggested that, on the one hand, resistance is low at any given point in a stream because internal feedbacks that can act to decouple the system from the immediate influence of external inputs are generally weak. On the other hand, stream resilience is high because each segment of stream is continually flushed out by new water, allowing rapid recovery of nutrient status and biomass levels following a perturbation.

These characteristics of streams make their response to watershed disturbances an interesting subject of study. Webster and Patten (1979) examined stream ecosystem response for three situations: (1) hardwood forest was removed, grass was planted, and natural succession was allowed to proceed; (2) hardwood vegetation was replaced by white pine; (3) a hardwood site was left undisturbed as a control. The nutrient dynamics (K and Ca) and food web characteristics of the stream were studied.

The results of the experiment showed generally higher nutrient levels in the streams of altered watersheds. A major food web change was the shift towards greater importance of crayfish and lesser importance of benthic insects in the perturbed streams. This shift was not explained, though a greater uniformity of input of nutrients through the year may have favoured the crayfish, which, unlike the univoltine benthic insects that have only one

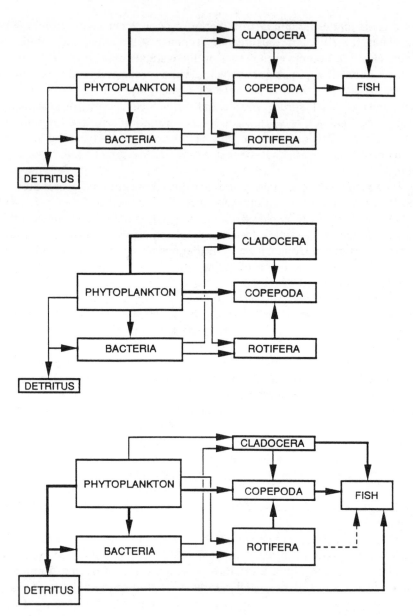

Figure 10.11 Changes in pelagic food web structure in Lake Gjersjoen caused by interactions between planktivorous fish, grazers and nutrient additions. The thickness of lines indicates the significance of an energy path. The three patterns are, from top to bottom; oligotrophic stable phase, eutrophic phase without fish predators and eutrophic phase after introduction of planktivorous/detritivorous fish. (From Hessen and Nilssen, 1986.)

generation per year, can produce several cohorts. The shift in the detrital community was accompanied by greater use of litterfall inputs, which may have been related to greater decomposability of litter inputs from the disturbed streams.

The results of the study also confirmed the resilience of stream systems. The perturbed systems were functionally similar to the unperturbed streams. They had not fundamentally changed and continued to differ only in ways attributable to different allochthonous inputs. Thus the streams seem to be limited in returning to their original states only by the rate of recovery of allochthonous inputs to their original values.

10.6 SUMMARY AND CONCLUSIONS

Variations in the environmental driving variables are natural occurrences in ecosystems and cause changes in the ecosystem as well as affecting the persistence of some species and communities. Variations in nutrient inputs and impacts to an ecosystem that indirectly affect nutrient inputs and fluxes are an important class of ecosystem perturbations.

The effect that an environmental change or pattern of variations has on a system depends not only on the characteristics of the change (intensity, duration, frequency, etc.) but also on the resistance and resilience characteristics of the system. In differential equation models of ecological systems, the eigenvalues of the equations are related to the resistance and resilience. Models with eigenvalues that are small in magnitude respond slowly to disturbances while models with large eigenvalues respond quickly and may track temporally varying driving forces.

The resistances to change of a variety of ecosystem types subjected to permanent changes in resource input were compared on the basis of the eigenvalues of linear nutrient flow models. The comparisons produced some expected results, such that stream ecosystems, with fast turnover rates, were much less resistant than ecosystems with large compartments (e.g. soil nutrients) that turn over slowly. However, less obvious patterns were also revealed, including significant changes in the resistance ranking of ecosystems depending on the temporal duration of the change.

Disturbances have been hypothesized to play a key role in maintaining species diversity in systems in which, under constant environmental conditions, competitive displacements would be expected to occur. Both model studies and controlled laboratory experiments have provided support for the hypothesis.

The behaviour of real ecological systems is usually non-linear, which creates the possibility of more than one stable equilibrium state for an ecosystem. A large enough perturbation away from one stable state can trigger positive feedbacks, including nutrient losses, that reinforce the perturbation and result in a new stable equilibrium. Ecosystems normally have negative feedback mechanisms that resist the loss of nutrients, but under stressed conditions perturbations can lead to positive feedback cycles. Examples of system level changes involving both initial nutrient losses (from forest systems) and additions (to aquatic systems) were discussed. Stream ecosystems, which tend to be low in resistance but high in resilience, may show significant shifts in community structure when subjected to a long-term disturbance, but appear to recover rapidly to their former state when the imposed change is removed.

11 Effects of spatial extent

11.1 INTRODUCTION

The study of interactions between nutrients and food webs is a complex matter even without taking into account the additional effects of movement in spatially heterogeneous landscapes. However, an understanding of the effects of the additional complexity resulting from spatial extent and heterogeneity is vital for understanding ecosystems.

Nutrients do not just stay in one spatial location and cycle endlessly through the food web, but also undergo horizontal and vertical movement through the ecosystem. This is obvious for such systems as streams and rivers, where the downstream transport of nutrients is one of the dominant processes in the ecosystem. However, lateral movement occurs in all systems since water flow occurs on even the smallest gradients. In the Everglades of southern Florida, for example, which has a south to north elevational gradient of only about 9 centimetres per kilometre, the slow movement of water from the agricultural area south of Lake Okeechobee is carrying nutrients from fertilizers southward. These nutrients flow into the northern parts of the Everglades, where the enrichment is causing changes in community composition from sawgrass to cattails (Davis, 1989). In moderately hilly areas such as the Iowan loess landscape, nitrogen is highly correlated with slope position, with nitrogen percentages at the bottom of slopes sometimes double those high on slopes (Aandahl, 1948; Jenny, 1980). In the ocean and lakes, nutrients are carried by currents from such sources as mouths of rivers and upwelling zones. The time scale of this horizontal movement may be slow or rapid, depending on the tightness of biological cycling of the nutrients with the stationary biota along the path of movement, the speed of movement of the medium (e.g. water) by which nutrients are transported through the system, and the rates of other processes of import and export, such as organism drift and directed animal movements.

In earlier chapters of this book movements in and out of the observed system were simply referred to as 'inputs' and 'outputs' of the system and no particular attention was paid to where the inputs came from or where the outputs went. Obtaining a grasp of the whole system means understanding how each spatial segment of the system affects every other spatial segment of the system. This requires that each input to a segment and each output from the segment be followed in its connections to other parts of the spatially extended ecosystem.

There are three basic ways in which nutrients move horizontally: (1) advection by directed movements of the medium, such as water flow and air currents, (2) diffusive movements, including molecular and turbulent diffusion of water and air, and (3) movements of organisms. Each of these modes of transport is discussed in this chapter along with the possible effects on food webs.

11.2 ADVECTIVE MOVEMENTS OF NUTRIENTS AND ORGANISMS

Advection means to be carried in a certain direction. Nutrients may be carried by directional flow of the water, either in dissolved or particulate form, where the particulate form may include both living and non-living particles. Nutrients may be carried by directional air currents as a gas, aerosol or small particles (e.g. soil, wind-blown organisms). Advective movements can change strength and direction temporally, as exemplified by tidal movements and seasonally changing wind or ocean water current patterns. Some advective movements may be permanently unidirectional, as is normally the case in streams and non tidally influenced rivers.

This section focuses on the type of unidirectional movement characteristic of streams. Even though permanent unidirectionality is simpler to deal with than advective movements that vary in direction with time, even fairly simple unidirectional transport of nutrients and biota and the interaction of this transport with ecosystem processes create complications that have made stream ecosystems difficult to grasp theoretically (Hutchinson, 1964; Webster, 1975). The linkages from upstream to downstream segments have such significance that it is difficult to consider these segments independently. Therefore, system boundaries on an axis along the direction of flow (longitudinal) are difficult to establish for model development.

Horizontal directional movement of nutrients via water flow is important not only in streams, but in most other ecosystems, including fens and bogs (Johnson, 1985), agricultural and forested land (Peterjohn and Correll, 1984) and coral reefs (D'Elia, 1988). The concepts derived explicitly for streams then have a wider applicability to many types of ecosystems. Below, the stream nutrient spiralling concept, based on unidirectional spatial transport, is described, as well as some of its more general implications for ecosystems.

The stream spiralling concept, which deals specifically with nutrient transport and spiralling, attempts to deal with the entire stream and river system rather than just small segments of it (Webster, 1975; Webster and Patten, 1979; O'Neill et al., 1979; Newbold et al., 1982; Elwood et al., 1983). The basic idea of nutrient spiralling is that when nutrient cycling is combined with advective movement through the system, a process of spiralling of nutrient molecules along the stream results. This is shown schematically in Figure 11.1, which pictures a reach of stream divided, for simplicity, into a water column and a layer of benthic biomass, such as a

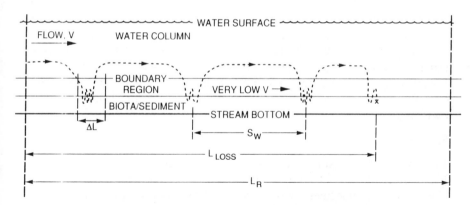

Figure 11.1 A schematic of a reach of stream of length L_R, showing three components, the flowing water column, a region of relatively stagnant water bounding the sediment and benthic biota, and the sediment and biota. A hypothetical nutrient atom (dotted line) enters the reach in the water column, drifts or diffuses into the boundary region a few times, where it cycles through the food web, and is finally lost (after moving some distance L_{loss}) through irreversible uptake by the sediment of conversion to a recalcitrant organic form. ΔL represents a small longitudinal distance the nutrient atom may move while recycling within the benthic region.

periphyton community. A nutrient molecule enters the upstream end of the water column and diffuses (via eddy or turbulent diffusion) into the benthic boundary zone, where water movement is slower and where most biomass is concentrated. There it may be taken up by an autotroph (or decomposer) and cycled a number of times through the food web before diffusing back into the water column and being carried farther downstream. This process can occur again and again along the stream, so that an atom of nutrient is used by the food web many times along the stream before being discharged into the ocean, permanently lost to sediments, removed from the stream by some process such as emergence of insects, or taken up by riparian plants via adventitious roots.

Newbold *et al.* (1981) showed that one could characterize the efficiency with which nutrients are used in a stream through a so-called spiralling length, or the average distance along the stream a nutrient atom travels per complete cycle. A complete cycle means 'biotic uptake of a nutrient atom from a dissolved available state, subsequent passage through the food chain, and ultimate return to the water in a dissolved form available for reutilization' (Newbold *et al.*, 1981, p. 629).

Suppose the stream ecosystem is in steady state and that it is longitudinally uniform. A nutrient molecule released into the water column upstream can be thought of as having an uptake rate per unit surface area of stream, U, which has units of $g\,m^{-2}\,s^{-1}$. This rate will depend on the density of autotroph biomass and its production rate (density of microbial biomass if bacterial and fungal immobilization of nutrients occurs, as it usually does in streams with large allochthonous nutrient inputs). If V_w is water velocity (m s^{-1}), N is the concentration of nutrient per unit volume of stream water (g m^{-3}), and h is the average water depth (m), then the expected distance of travel of the nutrient molecule before being taken up by biomass is

$$S_w = NhV_w/U \text{ (m)} \tag{11.1}$$

To understand this expression, think of a small sample cell of water of depth h and nutrient concentration N. The units of Nh are g m^{-2}. As it moves along, nutrient molecules are being removed by biomass uptake at a rate U (g m^{-2} s^{-1}). The mean time before a nutrient molecule is taken up is Nh/U (s). Since the cell is moving downstream at a velocity V_w, the mean distance travelled by the molecules is NhV_w/U. Hence, the greater the uptake rate is, the smaller the water velocity is, or the smaller the concentration of nutrients in the water are, the shorter the spiralling uptake length will be.

The uptake spiralling length S_w refers to nutrient transport downstream in solute molecule form. However, the nutrient may also be taken up by biota [only microbial uptake is assumed by Newbold *et al.* (1982) in their model, but this is generalizable to autotrophs also] and transported downstream in organic form (drifting living or detrital material) for varying distances before becoming available again. Let B be the concentration of nutrient in particulate form (g m^{-3}), R the release rate of available nutrients from the particulate form per unit time (g m^{-2} s^{-1}) and V_p (m s^{-1}) the average movement of the particulate forms downstream (this will generally be much smaller than V_w is, as particles will generally move more slowly on average than dissolved substances). Then the spiralling length in particulate form is, by analogy with (11.1),

$$S_b = BV_ph/R \tag{11.2}$$

The total spiralling length is, therefore,

$$S = S_w + S_b = NV_wh/U + BV_ph/R \tag{11.3}$$

Several things may be hypothesized relative to the idea of spiralling length:

(1) A nutrient-limited system may be expected to have evolved mechanisms to slow the movement of nutrient and decrease the spiralling length.
(2) The structure of the food web and the characteristics of the species in the food web may have a large effect on nutrient retention and spiralling length.
(3) Downstream areas will be affected by the retention upstream. This effect may be as important in dynamic situations, such as recovery from a disturbance, as it is in determining steady-state structure.

(4) The percentage of nutrient retention per length of stream may depend on the level of nutrient input, since high nutrient levels could saturate biotic uptake rates.

Calculations of spiralling length

The expression for spiralling length, S, derived above, is a function of the concentration of nutrient in solute form, N, and that in particulate form, B. However, B and N themselves depend on the process rates within the stream. Thus these quantities must be determined in order to determine S.

Newbold *et al.* (1982) calculated the way that total nutrient in the stream was partitioned into B and N by assuming that the total nutrient flux, F_T, was a constant consisting of both solute and particulate forms of the nutrient (I will use their notation NV_w here, rather than the notation for dissolved nutrient input, I_n, that I have used in earlier chapters);

$$F_T = NV_w + BV_p (g\,m^{-2}\,s^{-1}) \tag{11.4}$$

They also assumed expressions for nutrient uptake U and release rate R:

$$U = aBN/[(1 + k_1 N)(1 + k_2 B)] \tag{11.5}$$

$$R = rB \tag{11.6}$$

In steady state B and N will each be constant at a given location along the stream, so that uptake and release rates, U and R, are equal. Then, from Equations (11.5) and (11.6)

$$aBN/[(1 + k_1 N)(1 + k_2 B)] = rB \tag{11.7}$$

When Equations (11.4) and (11.7) are combined to eliminate N, an expression for B is obtained:

$$rk_1 k_2 V_p B^2 + (rk_1 V_p - rk_1 k_2 F_T - aV_p - rV_w k_2)B + aF_T - $$
$$-rV_w - rk_1 F_T = 0 \tag{11.8}$$

From this second-order equation in B, B can be computed and used in Equation (11.4) to determine N. It is then possible to determine the spiralling length S from Equation (11.3).

Newbold *et al.* (1982) were interested in how microbial self-limitation, represented by the coefficient k_2 in the expression for nutrient uptake (11.5), affects spiralling length. They first assumed no limitation, or $k_2 = 0$, for which situation B and N become

$$B = (F_T/V_p) - rV_w/[V_p(a - rk_1)] \tag{11.9}$$

$$N = r/(a - rk_1) \tag{11.10}$$

Now the spiralling length, S, can be calculated (for $k_2 = 0$):

$$S = S_w + S_b = (V_w h/aB)(1 + k_1 N) + V_p h/r$$
$$= (V_p h/r)\{1 + V_w/[F_T(a/r - k_1) - V_w]\} \tag{11.11}$$

Table 11.1   Parameter values for a stream ecosystem (Walker Branch, Oak Ridge, Tennessee) relevant to the calculation of the spiralling length

Symbol	Units	Value	Definition
a	$\mathrm{m^3\,g^{-1}\,s^{-1}}$	2.54×10^{-3}	Nutrient uptake rate coefficient [Equation (11.5)]
k_1	$\mathrm{m^3\,g^{-1}}$	200	Coefficient in Equation (11.5)
k_2	$\mathrm{m^3\,g^{-1}}$	0, 10, 50, 100	Microbial self-limitation coefficient [Equation (11.5)]
r	$\mathrm{s^{-1}}$	1.0×10^{-6}	Nutrient regeneration rate
V_w	$\mathrm{m\,s^{-1}}$		Downstream velocity of dissolved nutrients in stream
V_p	$\mathrm{m\,s^{-1}}$	23×10^{-6}	Downstream velocity of particulate compartment

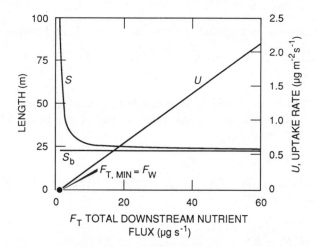

Figure 11.2   Model solutions for spiralling length, S, turnover length, S_b, and uptake rate of dissolved nutrient, U, as a function of total downstream flux, F_T, for the case of pure nutrient limitation ($k_2 = 0$); [Equation (11.11)]. (From Newbold *et al.*, 1982.)

Plotted as a function of total downstream flux, F_T, for the set of parameter values in Table 11.1, the spiralling length is shown in Figure 11.2. Spiralling length, S, approaches very large values when F_T decreases toward the limit $F_T = V_w/[(a/r) - k_1]$ because at that limit biota cannot survive and there is no uptake of nutrient from the water column. The spiralling length, therefore, approaches infinity.

Suppose next that microbial self-limitation takes on non-zero values. The solution for S is now more complex and will not be presented here, though

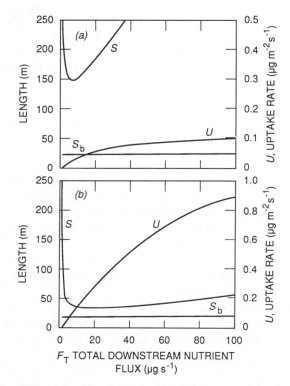

Figure 11.3 Model solutions as in Figure 11.2, but for cases of (a) strong microbial self-limitation $(k_2 = 100\,\text{m}^2\,\text{g}^{-1})$ and (b) weak microbial self-limitation $(k_2 = 10\,\text{m}^2\,\text{g}^{-1})$. (From Newbold *et al.*, 1982.)

the results will be presented. First let $k_2 = 100\,\text{m}^2\,\text{g}^{-1}$, so that self-limitation is strong. Now the spiralling length has more complex behaviour (Figure 11.3(a)). Because uptake rate U is limited, increasing values of total downstream nutrient flux F_T have little affect on U. However, the spiralling length S increases enormously as F_T increases beyond a certain level, because the amount of nutrient moving downstream overwhelms the ability of the microbial biota to take it up. When k_2 takes a lower value, representing weak microbial self-limitation, the results for S, S_b and U are intermediate between the cases of no self-limitation and high self-limitation (Figure 11.3(b)).

One can easily draw some biological conclusions of interest from the analysis of spiralling length. The values of some of the parameters in the expression for S are controlled by functional components of the food web. For example, it is well known that invertebrate grazers have positive effects both on the nutrient regeneration rate, r, and the downstream particulate velocity, V_p (Peterson and Cummins, 1974; Short and Maslin, 1977; Kitchell *et al.*, 1979). An increase in grazing intensity affects both V_p and r, so the resultant change in S can be complex. Preliminary estimates by Newbold *et al.* (1982) showed that for parameter values appropriate to Walker Branch stream in Oak Ridge, Tennessee, increases in the grazing rate would cause S to increase. These predictions were confirmed by experiments (Mulholland *et al.*, 1983).

11.3   DIFFUSIVE MOVEMENTS OF NUTRIENTS AND ORGANISMS

Diffusion is defined by Crank (1975) as 'the process by which matter is transported from one part of a system to another as a result of random molecular motions'. If a drop of red dye is carefully deposited with an eye dropper into a beaker of still clear water, the molecules of dye will gradually diffuse until the dye is spread uniformly through the water. Molecular diffusion can also occur in a stream of water that is flowing non-turbulently and at one velocity in some particular direction. In this case the individual molecules move in a random walk away from their centre of mass, which moves with the advective velocity.

The dynamics of the spread of concentration, $C(s,t)$, of a diffusing substance is described by a partial differential equation of diffusion. The simplest possible example of such an equation, for one dimension, is

$$\partial C(s,t)/\partial t = D\partial^2 C(s,t)/\partial s^2 \tag{11.12}$$

where D is the diffusion coefficient, assumed constant here, and s is the spatial dimension. The solution for this equation is

$$C(s,t) = (A/t^{1/2}) \exp(-s^2/4Dt) \tag{11.13}$$

where A is an arbitrary constant that can be fit to the initial concentration of diffusing molecules. (Note, however, that at time $t = 0$ the concentration becomes infinite according to this equation, so this model is too idealized to be a precise description of a real experiment.) Plots of $C(s,t)$ at three different times, t_1, t_2 and t_3, are shown in Figure 11.4.

The derivation of Equation (11.12) from simple ideas of random molecular motion has been described by Okubo (1980) and Berg (1983), among others. There are countless elaborations on this simple model to account for two- and three-dimensional diffusion, inhomogeneities in the fluid medium, fluid motions and other complexities. All diffusion models, however, rest on the assumption that the movements of particles are purely random, apart from the biases imposed on these movements imposed by the complexities just mentioned.

While the diffusion equation is strictly valid, at best, at the molecular level, where randomness can be assumed, analogues of this model have been used to describe movements on much larger scales and for particles of much greater than molecular size, such as biological organisms (Okubo, 1980; Berg, 1983).

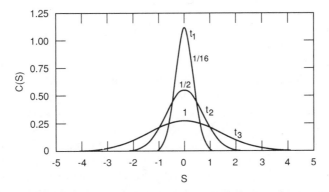

Figure 11.4   Solution of the diffusion equation (11.13) for $C(s, t)$ for three different times. (From Crank, 1975.)

Large-scale analogues to diffusion are created in water or air by turbulence (e.g. Csanady, 1973; Okubo 1980; Denny, 1988), which may be viewed as a superposition of eddies whose scales can reach macroscopic dimensions characteristic of the particular system.

The eddies in a turbulent system cause dissolved substances, as well as organisms floating in water or air, to spread out through time in a way that resembles molecular diffusion, though Okubo (1980) pointed out many ways in which turbulent diffusion in nature differs from the idealized model of molecular diffusion. Turbulent diffusion, in any case, occurs on spatial scales much greater than molecular diffusion and at much faster temporal scales than molecular diffusion could account for. Therefore, it is of most importance in ecological systems.

Other macroscopic factors besides turbulence of air and water media may create diffusion-like patterns. The basic movements of animals dispersing and searching for food may have some resemblance to diffusion and frequently have been modelled as diffusion or random walks. The ability of both nutrient and simple organisms to diffuse in aquatic and marine ecosystems has some interesting consequences for the stability of food chains, which are explored below.

Effects of organism diffusion that increase food chain stability

The importance of including space in models of ecological interactions, because of the way it can affect fundamental system behaviours, is an active area of ecological research. It was recognized that the ability of organisms to move and of nutrient molecules to diffuse could alter the stability characteristics of systems. After May (1973) and others showed, on purely mathematical grounds, that large complex ecological systems will tend to be unstable, various explanations were sought for the apparent stability of real ecological communities. Spatial considerations were an important part of many explanations. In reviewing some of the early work that attempted to take into account spatial heterogeneity and organism dispersal, Levin (1974, 1976) noted that these phenomena allowed competing species to coexist through dividing space, allowed extremely poor competitors but good dispersers to exist as 'fugitive' species, and allowed prey to avoid being overexploited by predators through predator avoidance in space. It was shown, in particular, that diffusion in space could in some cases smooth out instabilities such as those resulting from strong interactions between predators and prey (Gilpin, 1975). A recent review of the literature (DeAngelis and Waterhouse, 1987) substantiates the prevalent opinion among ecologists that spatial extent and heterogeneity, along with varying abilities of organism species to disperse, plays a vital role in the diversity and relative stability of ecological communities.

The examples below indicate that the spatial diffusion of organisms can actually have mixed effects on ecosystem stability, sometimes helping to reduce incipient instabilities, but sometimes creating conditions for instability.

Steele (1974a) analysed a model of plants and herbivores (phytoplankton and zooplankton) in a marine pelagic environment to study the effects of including diffusion. Sjoberg (1977) followed up on this by including the effects of nutrient limitations. The equations describing this system are as follows (where some notational changes have been made):

$$\partial N_x/\partial t = I_n + f_4(N_y) - f_1(N_x)N_y + D\partial^2 N_x/\partial s^2 \qquad (11.14a)$$

$$\partial N_y/\partial t = f_2(N_x)N_y - f_3(N_y) + D\partial^2 N_y/\partial s^2 \qquad (11.14b)$$

where N_x and N_y represent phytoplankton and zooplankton densities, or, more precisely, the limiting nutrients bound up in the phytoplankton and zooplankton, along one spatial dimension, s. The term I_n is a constant representing the part of the growth rate of phytoplankton permitted by the input of new limiting nutrient into the system, while $f_4(N_y)$ represents the part of the growth rate allowed by recycling of nutrient that has passed through the zooplankton component. For simplicity, this recycling is assumed to occur instantaneously, without the nutrient atoms having to pass through a stage of being stored in an available pool of dissolved nutrient.

The term $f_3(N_y)$ [where $f_4(N_y) = k_1 f_3(N_y)$, $0 < k_1 < 1$] is the zooplankton loss rate, the fraction $1 - k_1$ of which sinks out of the pelagic system. The term $f_1(N_x)N_y$ is the loss rate of phytoplankton nutrient, where an amount $f_2(N_x)N_y$ [$f_2(N_x) = k_2 f_1(N_x)$; $0 < k_2 < 1$] reaches the zooplankton and the rest sinks out of the system.

No specific functional forms were given for $f_1(N_x)$ or $f_3(N_y)$, except that they were assumed to increase with increasing N_x and N_y respectively.

Sjoberg (1977) defined a steady-state equilibrium point (N_x^*, N_y^*). Then he linearized Equations (11.14a,b) around this point to look at the stability characteristics in a way that has been described in earlier chapters. Letting $N_x = N_x^* + N_x'$ and $N_y = N_y^* + N_y'$, where $N_x' \ll N_x^*$ and $N_y' \ll N_y^*$, he obtained,

$$\partial N_x'/\partial t = a_{11}N_x' + a_{12}N_y' + D\partial^2 N_x'/\delta s^2 \tag{11.15a}$$

$$\partial N_y'/\partial t = a_{21}N_x' + a_{22}N_y' + D\partial^2 N_y'/\partial s^2 \tag{11.15b}$$

where

$$a_{11} = -[df_1/dN_x^*]N_y^* \tag{11.16a}$$

$$a_{12} = k_1[df_3/dN_y^*] - [k_1 f_3 N_y^*/N_y^*] - I_n/N_y^* \tag{11.16b}$$

$$a_{21} = k_2[df_1/dN_x^*]N_y^* \tag{11.16c}$$

$$a_{22} = f_3(N_y^*)/N_y^* - df_3/dN_y^* \tag{11.16d}$$

If diffusion is ignored, by letting $D = 0$, then, from an eigenvalue analysis similar to that done in earlier chapters [letting $N_x' = \exp(\lambda t)$ and $N_y' = \exp(\lambda t)$ in Equations (11.15a,b) and examining the sign of the λ values in the resultant equation], the necessary and sufficient conditions for stability are

$$a_{11} + a_{22} < 0 \tag{11.17a}$$

$$a_{11}a_{22} - a_{12}a_{21} > 0 \tag{11.17b}$$

(Proof of these conditions is left to the reader.) Because $f_1(N_x)$ and $f_3(N_y)$ are increasing functions, then $df_1/dN_x^* > 0$ and $df_3/dN_y^* > 0$, so that $a_{11} < 0$ and $a_{21} > 0$. If $a_{22} < 0$, or

$$f_3(N_y^*)/N_y^* - df_3/dN_y^* < 0 \tag{11.18}$$

then (11.17a,b) are automatically satisfied. The reasoning is as follows. Since a_{11} is always negative, (11.17a) is true. Substitution of (11.16a,b,c,d) into (11.17b) also shows that this inequality is satisfied. The system is, therefore, stable.

How does diffusion affect this situation? By letting $D > 0$, one takes into account the spatial aspect of possible instabilities by letting the spatial variation of N_x' and N_y' be decomposed into a Fourier series of components. What this means is that the variations in space of a variable representing a population is imagined to be represented as a series of sinusoidal curves (see Okubo, 1978, for example), a substitution that allows analytical treatment:

$$N_x' = \sum_{\xi} a_\xi \exp(\lambda t + i\xi s) \tag{11.19a}$$

$$N_y' = \sum_{\xi} b_\xi \exp(\lambda t + i\xi s) \tag{11.19b}$$

where i indicates an imaginary number and ξ is a particular frequency of sinusoidal change along one spatial dimension, measured by s. Then the inequalities necessary and sufficient for stability, (11.17a,b), are replaced by

$$\hat{a}_{11} + \hat{a}_{22} < 0 \tag{11.20a}$$
$$\hat{a}_{11}\hat{a}_{22} - \hat{a}_{12}\hat{a}_{21} > 0 \tag{11.20b}$$

where

$$\hat{a}_{11} = a_{11} - D\xi^2 \tag{11.21a}$$
$$\hat{a}_{22} = a_{22} - D\xi^2 \tag{11.21b}$$

One can examine (11.20a,b) in the same way as (11.17a,b) were examined. Since diffusion causes both $\hat{a}_{11}$ and $\hat{a}_{22}$ to become more negative, and hence $\hat{a}_{11}\hat{a}_{22}$ to become more positive, for any spatial mode ξ (where $1/\xi$ represents the spatial scale of any incipient unstable fluctuation), the addition of diffusion increases the range of parameters for which the inequalities (11.20a,b) are satisfied and the system is stable. Even if, contrary to the assumption (11.18), $a_{22} > 0$, a sufficiently large value of D can stabilize the system against smaller spatial scale modes of fluctuation (i.e. against modes for which ξ is large enough that $a_{22} - D\xi^2 < 0$).

Diffusion effects that decrease food chain stability

Okubo (1974, 1978) also considered a nutrient–phytoplankton–zooplankton system, but it differed in details from that of Sjoberg (1977). The nutrient concentration was modelled explicitly, rather than implicitly as in Sjoberg's model, and the zooplankton population was not included as an explicit variable, but only through its effects as a constant harvesting rate on the phytoplankton.

The model equations were (with some changes in notation)

$$\partial N/\partial t = I_n - f_1(N)N_x + D_1\partial^2 N/\partial s^2 \tag{11.22a}$$
$$\partial N_x/\partial t = -d_2 + f_2(N)N_x + D_2\partial^2 N_x/\partial s^2 \tag{11.22b}$$

where I_n is the input of nutrient and d_2 is the constant grazing rate (note that this is independent of phytoplankton standing stock). In this model the diffusion rates, D_1 and D_2, can be different from each other.

When the equations are linearized about the equilibrium point (N^*, N_x^*) to examine stability, by substituting $N = N^* + N'$ and $N_x = N_x^* + N_x'$, the resultant equations are

$$\partial N'/\partial t = a_{11}N' + a_{12}N_x' + D_1\partial^2 N'/\partial s^2 \tag{11.23a}$$
$$\partial N_x'/\partial t = a_{21}N' + a_{22}N_x' + D_2\partial^2 N_x'/\partial s^2 \tag{11.23b}$$

where

$$a_{11} = -(df_1/dN^*)N_x^* < 0 \tag{11.24a}$$
$$a_{12} = -I_n/N_x^* < 0 \tag{11.24b}$$

$$a_{21} = (df_2/dN^*)N_x^* > 0 \tag{11.24c}$$
$$a_{22} = d_2/N_x^* > 0 \tag{11.24d}$$

and where the inequalities result from the fact that $df_1/dN^* > 0$ and $df_2/dN^* > 0$.

When substitutions analogous to (11.19a,b) are made in Equations (11.23a,b), the inequalities (11.20a,b) again result as necessary and sufficient conditions for stability. In this case, these inequalities are

$$df_1/dN^*N_x^* + d_2/N_x^* - \xi^2(D_1 + D_2) < 0 \tag{11.25a}$$
$$[df_1/dN^*N_x^* + \xi^2 D_1][(d_2/N_x^*) - \xi^2 D_2]$$
$$+ (I_n/N_x^*)[(df_2/dN^*)N_x^*] > 0 \tag{11.25b}$$

Suppose that, even when $D_1 = D_2 = 0$, inequality (11.25a) is satisfied; i.e. $f_1(N^*)N_x^* > d_2/N_x^*$. Then adding diffusion only strengthens that inequality by making the left-hand side of (11.25a) more negative. However, if D_1 is large enough, inequality (11.25b) may be reversed for some values of ξ. Suppose that for $\xi = 0$ the inequality (11.25b) holds; that is, the second term is larger in absolute magnitude than the first one, which is negative. As ξ increases, the first factor of the first term increases in magnitude, whereas the second factor decreases (as the $\xi^2 D_2$ term increases relative to the d_2/N_x^* term). If D_1 is large enough and D_2 small enough, then the first term will stay negative and as a whole will increase in absolute magnitude as ξ is increased to moderately large values. Eventually, the first term will exceed the second term in magnitude and the system will be unstable. Thus, when $D_2 \ll D_1$ it is possible for medium spatial scale (ξ moderately large) instabilities to develop. If ξ is very large, such that $\xi^2 D_2 > d_2/N_x^*$, the sign of the left-hand side reverses again, so the system is stable to very small spatial scale instabilities (ξ very large).

Under what circumstances might D_2 be much smaller than D_1, meaning that nutrients diffuse faster through the water than phytoplankton do? Okubo (1968) noted that phytoplankton concentrate near the water surface and nutrients are spread nearly uniformly in the upper mixed layer of the ocean. Because a major source of diffusion, shear diffusion, increases with depth, this could result in phytoplankton dispersion being smaller than the dispersion of dissolved substances by an order of magnitude. In particular, the value of ξ^2 that causes (11.25b) to reach the switching point [where the left-hand side of (11.25b) becomes zero] is

$$\xi^2 = -\tfrac{1}{2}[(df_1/dN^*)(N_x^*/D_1) - d_2/(N_x^* D_2)] \pm \tfrac{1}{2}\{[(df_1/dN^*)(N_x^*/D_1)$$
$$- d_2/(N_x^* D_2)] - 4[(I_n df_2/dN^*) - (d_2 df_1/dN^*)]/(D_1 D_2)\}^{1/2} \tag{11.26}$$

Explanation of diffusion effects

The results concerning diffusional effects on biotic stability derived above are mathematical and they may seem somewhat puzzling from an intuitive point of view. In the first example, Sjoberg's model, the effect of diffusion was to counteract any tendency towards instability, whereas in the second example, Okubo's model, diffusion could promote instability. How can these results be reconciled in biological and physical terms?

Sjoberg's model makes more intuitive sense and so is in less need of explanation. Instability in his model could feasibly occur from zooplankton overexploitation of phytoplankton. For the parameter values used in the model, however, the system in the absence of diffusion ($D_1 = D_2 = 0$) was stable. When diffusion was included, the system became even more strongly stable. Any incipient fluctuation towards zooplankton overexploitation of prey in one region of the spatially extended system would be counteracted not only by the stable nature of the predator–prey interaction, but also by diffusion; that is, a local upward fluctuation in zooplankton biomass would tend to diffuse away into the surrounding volume and the corresponding downward fluctuation in phytoplankton biomass would be compensated for by diffusion of phytoplankton in from the surrounding volume [see also Powell and Richerson (1985)].

Why does this mechanism of diffusional smoothing not work in Okubo's example? The potential destabilizing element in this case is the positive a_{22} term (d_2/N_x^*), which represents self-amplification of the phytoplankton; that is, since the effect of zooplankton on the phytoplankton is simply at a constant harvesting rate, d_2, that does not change through time, any increase in phytoplankton density does not call into play increased regulation from the zooplankton trophic level. The counteracting stabilizing effect in the model is the decrease in nutrient concentration that comes from any upward fluctuation in phytoplankton biomass and that tends to pull the phytoplankton level back towards the equilibrium level. However, diffusion of nutrient can now play a role in reinforcing an upward fluctuation of phytoplankton in some region of space – if only the nutrient diffuses rapidly and not the phytoplankton. The incipient phytoplankton upward fluctuation will continue to grow in the locale it started in because the phytoplankton will not diffuse away on short time scales, but nutrient from the surrounding volume can quickly diffuse in to the locale. Meanwhile, as a phytoplankton 'bloom' developed in this locale, phytoplankton numbers become depressed in surrounding areas because of nutrient depletion through diffusion. This results in spatial patches of phytoplankton that appear and disappear in time. The spatial scale of the patches is of the order of the values of $(1/\xi)$ for values of ξ that cause inequality (11.25b) to be reversed.

11.4   ORGANISM MOVEMENTS

For organisms on land the transport of material elements by the gravitational movement of water towards the oceans would seem to pose serious problems. Critical nutrients not significantly present in the atmosphere, such as phosphorus, would eventually be depleted if these generally unidirectional fluxes were not replenished by some means.

Living organisms play a role in the replenishment of nutrients. The role of biota in producing available nutrients in a given locale has been very well described by Hutchinson (1948) for a lake:

'Starting with a barren glacial basin newly filled with water, the concentration of phosphate in solution in the drainage basin will depend on the general geochemistry of the region. At first, the nitrogen available will probably be derived solely from rain water, but nitrogen will tend to be fixed biologically according to the availability of organic matter, the production of which will depend on the availability of phosphorus and other nutrients. The water of the lake will gradually develop a phytoplankton population. The first sediments to be deposited will be almost entirely inorganic, but as soon as remains of organisms are included in the surface layer of these sediments, an internal cycle of the kind already described will be established. Phosphate will be more easily liberated from mineral particles, owing to the production of carbon dioxide by decomposition, and the phosphate of the decomposing organisms will be returned to the lake water. As productivity increases, the sediments will become more and more organic, and thus more and more able to return the nutrients rapidly to the cycle. The process continues until the geochemically determined nutrient potential of the silt and the water of the drainage basin is fully utilized.'

Organisms not only create local fluxes of nutrients from rock to the overlying water or soil, but also can transport and redistribute nutrients over long distances. Juday *et al.* (1932) observed that the Pacific salmon may carry significant amounts of phosphorus to their upstream spawning areas, which they contribute to the stream ecosystem when they die after spawning. How important is this contribution? Donaldson (1967) and Krokhin (1967) demonstrated the importance of upstream migration of adult salmon. In a heavy spawning year, the amount of phosphorus put into the system by salmon carcasses equalled or exceeded inputs from other sources. This could be important in the growth and survival of larval and juvenile salmon (Parsons *et al.*, 1970). In other cases, fish migration in inland waters may not be as important. Hall (1972) investigated the phosphorus budget of a small stream in North Carolina, in which there was migration upstream of large fish. Measurements made over the course of a year of phosphorus movement in water, leaves and fish indicated that phosphorus movement downstream in water (solution and suspension) was by far the dominant flux. The net amount of phosphorus brought upstream by fish in the stream was less than 0.2% of that lost through stream discharge.

Another example of local nutrient enrichment via animal movements can occur at colonial bird nesting sites. Studying an abandoned bird rookery in the Okeefenokee Swamp, Oliver and Legovic (1987) noted indications both of local nutrient increases and of increased biomasses of various trophic levels. For example, equilibrium concentrations of benthic detritus were 3.7 times as great in the abandoned rookery as in a control site. A simulation model calculated that fish biomass in the rookery was about 1.42 times that in the control site. Thus the bird rookery helps, through nutrient import, to increase its local food source.

Mammals also play important roles in altering spatial distributions of nutrients. Deposition of urine and faeces around burrows can create high local concentrations and affect local vegetation structure (Woodmansee, 1978). Gophers, however, have been shown to reduce nitrogen in surface soils by creating mounds of subsurface soil (Inouye *et al.*, 1987; Koide *et al.*, 1987). This can increase landscape heterogeneity.

In these salmon, colonial bird, mammal and many others, higher trophic level species play a major role in redistributing nutrients and, therefore, in affecting the base of the food web in certain areas.

11.5 SUMMARY AND CONCLUSIONS

Spatial extent and the differential movement of nutrients and organisms in space complicates the description of the interaction of nutrients and food webs. Nutrient uptake or release in one location inevitably has effects elsewhere as a result of nutrient and organism movement. Two general modes of nutrient transport between different spatial regions of an ecosystem have been studied systematically: advective flow, such as occurs in streams, and diffusive movement in more lentic bodies of water. A third type of nutrient flux, movement via animal migrations and congregations, has also been recognized as being important in some situations.

One important idea relating to the advective movement of nutrients in streams has been termed 'nutrient spiralling', which refers to the fact that a given atom of nutrient may be taken up from water and utilized in the food web many times in the course of movement downstream, thus delaying its transport downstream. In this way, a nutrient atom may be used and reutilized. There is a characteristic scaling length, the spiralling length, associated with this phenomenon, and defined as the mean distance an atom travels from release into the water column until it is taken up biotically and released again into the water column. The spiralling length can be calculated in terms of basic physical and biological parameters of the stream and may be useful as a general indicator of stream dynamics. Knowledge of this characteristic length can help one predict how an upstream disturbance (perturbation of the nutrient budget or of the biota) will affect downstream components.

Nutrient and organism diffusion have been studied and modelled primarily with respect to their roles in affecting the stability of pelagic food webs, especially predator–prey relationships between zooplankton and phytoplankton. Diffusion generally plays a smoothing role, tending to damp out any incipient instabilities. However, under certain circumstances, when the rate of diffusion of two components of the system are widely different, diffusion can actually promote a tendency towards instability.

Animal movements may in some cases play a significant role in reversing the normal downhill movement of nutrients, such as by the restoration of some of the nutrients to upstream spawning areas by migrating fish or the creation of local concentrations of nutrients by colonies of nesting birds or burrowing mammals. Nutrient redistribution in space can influence the dynamics of food webs in significant ways.

12 Implications for global change

12.1 INTRODUCTION

Chapter 1 described, on a global scale, the material fluxes driven by energy from the sun. In subsequent chapters the focus narrowed to the ecosystem scale, or to areas small enough to be relatively homogeneous but large enough to be regarded as self-sustaining ecological units, except for possible inputs and losses of nutrients. A variety of phenomena that can be mechanisms involved in environmental problems at the ecosystem scale were touched on in earlier chapters. Several of these are listed in Table 12.1.

It is time now to return to the global scale to learn that human influence is not confined merely to local ecosystems, but extends to the planet as a whole. The burning of fossil fuels has resulted in an increase in carbon dioxide (CO_2) in the atmosphere, which poses the prospect of global

Table 12.1 Various environmental mechanisms involving nutrients in food webs that can be involved in environmental problems of importance to human society. The book chapter in which they are discussed is noted

Mechanism	Chapter
Resistance of whole lake to changes in nutrient input	2
Resilience of ecological systems to disturbances	2,3,4,5,6,7,8
Catastrophic die-offs of forest stands due to toxic effect of a nutrient	4
Autotroph–herbivore instabilities triggered by nutrient enrichment	5,6,7
Insect outbreaks triggered by nutrient enrichment	5
Catastrophic changes in ecosystems related to changes in nutrient level; e.g. eutrophication	8
Competitive displacement related to nutrient enrichment	9
Resistance of food chains to press disturbances of nutrient flows	10
Irreversible changes in ecosystems due to large disturbances	10
Effects of organism diffusion on food chain stability	11

changes in climate. This is a cause for great concern, because of the possible effects of climate change on, among other things, our water and food resources. The possibility of 'greenhouse' warming of the earth presents humankind with one of its most serious environmental challenges.

The set of problems associated with increased atmospheric CO_2 and the greenhouse effect on climate are enormously diverse and complex, having social, economic, and political components as well as scientific ones involving climatology, oceanography and ecology. Perhaps the most disturbing aspect of these problems, even the purely scientific ones, is that it is impossible to predict with any accuracy how increased atmospheric CO_2 is going to affect climate or even how much of the CO_2 released by fossil fuel burning will remain in the atmosphere. At the heart of the scientific dilemma are questions of how the large-scale biotic systems of the plant will respond to additional inputs of a major nutrient, carbon, how other nutrients affect this response, and how the biotic systems will change if climatic change occurs. I will review some of the factors involved in these key questions and indicate how models are being used to address them.

The CO_2 greenhouse warming issue can be divided into three general categories: (1) the human effects on the global carbon cycle, (2) the effects of increasing atmospheric CO_2 on climate and (3) the effects of climatic changes on the environment (including human society and the economy). Many questions of particular interest from the point of view of nutrient cycling in the biosphere fall under the first and third categories, though there is space here to consider only a few of these.

12.2 THE GLOBAL CARBON CYCLE

Measurements taken at Mauna Loa Observatory have revealed a steady increase in atmospheric CO_2 from about 315 to over 340 parts per million by volume (ppmv) since 1958 (Bacastow and Keeling, 1981). It has been estimated, from measurements of air bubbles in polar ice, that the present CO_2 level has risen from about 280 ppmv at the beginning of the nineteenth century (Neftel et al., 1985).

This increase in atmospheric CO_2 has been largely attributed to input from fossil fuel burning since the beginning of the Industrial Revolution. However, the measured increase in atmospheric CO_2 accounts for only about 58% of the CO_2 released by this source (Trabalka et al., 1985). To explain this discrepancy the terrestrial or oceanic component of the biosphere must be invoked as a sink. This explanation is confounded, however, by calculations, based on estimates of clearing of forests for agriculture, indicating that the terrestrial biota have been a net source of CO_2 to the atmosphere rather than a sink (Woodwell et al., 1983; Houghton et al., 1985; Detwiler et al., 1985).

The ocean is the most likely sink for the 'missing' CO_2, but current models of ocean–atmosphere interchange, which incorporate both vertical and lateral water movements, suggest that the ocean can take up to at best 35% of the fossil fuel released between 1958 and 1980 (Broecker *et al.*, 1986; Peng, 1986).

Thus, current understanding of the fluxes of carbon between the atmosphere, ocean and land cannot explain the observed levels of CO_2 in the atmosphere. Several critical gaps in knowledge have been identified that may need to be filled before an appreciable improvement in understanding of the global carbon cycle will be achieved (Post *et al.*, 1990). Most important among these are two processes that involve nutrient–food web interactions: (1) the fertilization effect of CO_2 on terrestrial systems that could increase carbon storage in biomass and soil, and (2) the role of biological productivity in the ocean in converting inorganic carbon to organic detritus that sinks to the deep.

Both of these problems relate to possible stimulating effects of carbon (CO_2) on primary productivity, contrary to earlier assertions (e.g. Table 3.2) that nitrogen, phosphorus and occasionally potassium, calcium or silicon are most frequently limiting. There are a few general reasons why increases in CO_2 may stimulate increased production. First, Liebig's Law of the Minimum (discussed in Chapter 3) is strictly true, if at all, only at steady-state equilibrium. In real situations, ambient levels of nutrients fluctuate and nutrient requirements of autotrophs change through time. More than one nutrient may be in the position of a limiter over some period. Second, increased CO_2 may sometimes affect a plant by allowing it to keep its stomata closed longer to conserve water. Less water transpiration might decrease the uptake of some mineral nutrients, such as nitrogen. Thus a simultaneous increase in CO_2 and some nutrients in the soil might have a larger positive effect on plant growth than an increase in either of these separately.

Effects of CO_2 enrichment on terrestrial plants

If carbon in the form of CO_2 acts as a limiting nutrient for terrestrial plant production, then increases in atmospheric CO_2 should induce a faster growth rate for the autotrophic component of the biosphere. Short-term experiments with annual species that use the C_3 photosynthetic pathway show that growth and productivity increase by 30–50% with a doubling of CO_2 in growth chamber environments with no other constraints (Strain and Cure, 1985).

To follow up on the implications of this idea theoretically, one can use the type of mathematical model formulated for the atmospheric–terrestrial components of the global carbon cycle by Ericksson and Welander (1956) (they considered the effects of the ocean in a more general version of the model). The authors were interested in the effects of non-linearities in the primary production and organic matter decomposition processes on behaviour of the global system. This model includes only atmospheric carbon

(CO_2), denoted by N here, carbon in living plants, N_x, and carbon in dead organic matter, N_D. A version of this model, simplified to consider only the non-linearity related to the fertilization effect, is

$$dN/dt = -r_1(N, N_x)N_x + d_1 N_x + d_D N_D \tag{12.1a}$$

$$dN_x/dt = r_1(N, N_x)N_x - d_1 N_x - d_2 N_x \tag{12.1b}$$

$$dN_D/dt = d_2 N_x - d_D N_D \tag{12.1c}$$

No increase in the decomposition rate coefficient for plants with higher plant biomass is assumed. To facilitate analysis this model ignores oceanic carbon and assumes that the total carbon in the system is

$$N_T = N + N_x + N_D \tag{12.2}$$

This total carbon increases through time because of fossil fuel input. The fertilization effect on plant growth, represented by $r_1(N, N_x)$, is assumed to be an increasing function of N for small values of N, but to decrease as N becomes very large, because atmospheric concentrations of CO_2 close to 1% may be inhibiting or even lethal to plants (Ballard, 1941). High levels of N_x are also assumed to inhibit plant growth because of plant crowding and competition.

The rough qualitative shapes of the zero isoclines, $dN/dt = 0$ and $dN_x/dt = 0$, after elimination of N_D from Equation (12.1a) by means of (12.2), are shown in Figure 12.1(a). There are three equilibrium points for this system, two of which, E_1 and E_3 (where the latter is an equilibrium for no living biomass), are stable, whereas E_2 is unstable. The dashed line shows a feasible trajectory. As N_T increases, the $dN/dt = 0$ isocline shifts upwards (Figure 12.1b). The amount at which the steady-state equilibrium value of atmospheric CO_2, N^*, increases for a given increase in N_T depends on the slope of the $dN_x/dt = 0$ isocline at E_1. As long as a large fertilization effect occurs, then N^* should increase much less rapidly than N_T. But it should be stressed that this applies only for the steady state, N^*. It is actually unlikely that the global carbon cycle is close to equilibrium now or that it will be for many years, even if fossil fuel burning and other disturbances were to stop now. The turnover time of carbon in the terrestrial vegetation is of the order of several decades and for the ocean as a whole is several hundred years. Thus, there is no question of the atmospheric CO_2 being in equilibrium with a rapid input of fossil CO_2.

Nevertheless, over long time periods, the fertilization effect, if it is as simple as is shown in Figure 12.1, could be a stable mechanism for buffering the atmospheric CO_2 increase. However, tests on fertilization effects have largely been done in growth chambers and on herbaceous plants, which constitute less than 20% of the global terrestrial living biomass (Olson *et al.*, 1983). More helpful would be long-term data on the effects of fertilization on forest communities, and these data do not currently exist. Only such studies could show if increased CO_2 concentrations would positively

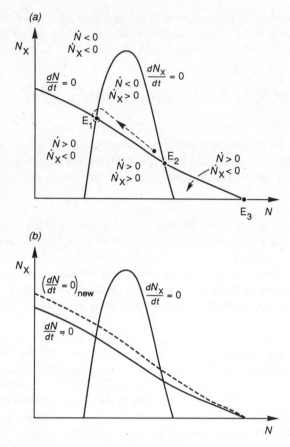

Figure 12.1 (*a*) Zero isoclines $dN/dt = 0$ and $dN_x/dt = 0$ for a model of atmos-pheric–terrestrial plant global carbon interactions. E_1, E_2 and E_3 are equilibrium points. N and N_x represent time derivatives of N and N_x. The dashed line represents a feasible trajectory. (*b*) Shift in the dN/dt isocline that occurs when total carbon in the system, N_T, increases. (Based on model by Ericksson and Welander, 1956.)

affect terrestrial carbon storage, or if bounds on plant growth, due to limitations of other nutrients, water and space, or increased herbivory would limit this effect. Long-term experiments **have** been performed on tundra plants under field conditions and these show no increase in carbon seques-tering with an increase in ambient CO_2 (Oechel and Strain, 1985).

Oceanic production: The biological pump

Analogous to the hypothesized fertilization of terrestrial plants by addi-tional input of CO_2 to the global system is its possible stimulation of productivity of marine phytoplankton. Such an increase in phytoplankton production could increase the amount of carbon transported to the deep oceans (and thereby effectively stored for a long time) by the sinking of organic detritus. This mechanism of carbon transport is sometimes referred to as the 'biological pump'.

Toggweiler *et al.* (1987) developed a model for a marine pelagic system that can simulate observed primary productivity, surface chlorophyll, surface nutrients and detrital sinking fluxes in some areas of the ocean to which it has been applied. The model contains compartments for phytoplankton, limiting nutrient (NO_3^-), suspended detritus, sinking detritus particles and two classes of grazers (Figure 12.2). The grazers feed on phytoplankton and produce faecal pellets that sink from the euphotic zone. This loss of organic nitrogen to the deep ocean is compensated for by a return flux of inorganic NO_3^- from ocean layers below the euphotic zone by means of eddy diffusion.

Toggweiler *et al.*, (1987) found it useful to include two grazer classes, large and small, in order to be able to simulate observed pelagic conditions. The small grazers were the primary consumers of the phytoplankton. The large grazers (resembling copepods) fed on both smaller grazers and phytoplankton, though their feeding efficiency on the latter was not as high as that of the small grazers. The large grazers egested large faecal pellets, which sank, whereas the faecal pellets of the small grazers remained suspended and recycled nutrients back to the water.

Widely different fluxes could result depending on the relative secondary production of the small and large grazers. 'When the large grazers are barely viable the primary production is very high because nutrients are extensively recycled. When the small grazers are barely viable, the primary production is low because a greater fraction of the primary production is incorporated into sinking faecal particles which remove nutrients from the upper layers' (Toggweiler *et al.*, 1987). The presence of a high concentration of small grazers relative to large ones held down algal productivity in systems where

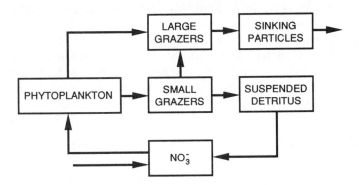

Figure 12.2 Schematic of model used by Toggweiler *et al.* (1987) to explore transport of organic carbon to deep ocean waters through the sinking of detrital particles.

the nutrient concentration in the euphotic zone was high, limiting the amount of organic carbon and organic nitrogen 'pumped' out of the system by the sinking of faecal pellets. The model was useful in showing that the system behaviour is very sensitive to the parameters determining the relative success of the two grazer types, as well as being sensitive to the influx of new nutrients, suggesting that these quantities be carefully monitored in the field.

The model was used to determine the effect of higher nutrient inputs from lower ocean layers. Increases in nutrient input did not increase phytoplankton populations, at least up until very high inputs, but resulted in larger populations of large grazers and higher fluxes of sinking faecal pellets. (This consumer regulation of the autotroph resembles models of Chapter 5.) Comparisons of the modelled flux of faecal pellets with that found in sediment traps, however, showed that these fluxes are much smaller than predicted, suggesting to the authors that much of the nitrogen flux out of the euphotic zone may be in the form of recalcitrant dissolved organic nitrogen rather than in detritus.

The model of Toggweiler et al. (1987) is only a simple approximation of upper ocean ecology and has been fitted to data only in certain parts of the ocean (the North Atlantic sub-tropical gyre near Bermuda and the subarctic North Pacific). Highly resolved three-dimensional models with more greater biological detail may be necessary to describe the entire ocean component accurately.

One important fact from the point of view of the ocean's capacity to buffer increased CO_2 to the atmosphere is that the flux of carbon out of the euphotic zone to deep ocean water must be equivalent in a stoichiometric sense to the supply of nutrient, usually NO_3^-, from what would limit primary production if inorganic carbon is present in abundance. Thus far, estimates of the supply of nitrate brought to the euphotic zone by vertical turbulent diffusion from deeper water is not sufficient to account for the estimates of primary production based on measurements of oxygen consumption (Lewis et al., 1986). Thus better determination of this input flux is vital.

Another interesting aspect of the biological pumping idea is the recent suggestion that phytoplankton in the otherwise nutrient-rich Antarctic oceans are limited by iron rather than nitrate or phosphate (Martin et al., 1990). Antarctica is farther from land sources that supply wind-carried iron to other oceanic areas.

12.3   EFFECTS OF CLIMATIC CHANGE ON THE BIOSPHERE

If increased greenhouse warming from CO_2 and other gases affects the climate, then the earth's ecosystems will also change in response. Forecasts of what changes will occur under a given climate scenario would be highly valuable in planning. Climate models suggest that a doubling of atmospheric CO_2 from the pre-Industrial Age level, which may occur during the twenty-first century, could cause a mean rise in global temperature in the range 2–4°C. Changes in precipitation and atmospheric and oceanographic circulation are also expected to occur (Schneider, 1989). A few of the many possible effects that relate to food webs and nutrient cycles are discussed briefly below.

Effects on forest ecosystems

Pastor and Post (1988) considered the effects of a warmer drier climate at the boundary of the boreal forest and cool temperate forest in northeastern North America. They found that the forest response due to altered ecosystem feedbacks resulting from predicted climate changes will be larger than the responses predicted purely from physiological responses of the plants, at least over long time scales of 200 years or so. At these scales feedback loops involving population dynamics, competition, succession and plant–soil interactions have time to act and to constrain the potential direct physiological responses (Figure 12.3). The forest environment, the species composition and the associated soil properties all change because of the feedback loops linking these properties.

The approach used by Pastor and Post (1988) to predict change was an individual-based forest simulation model; that is, trees of all relevant species were modelled as individuals growing and competing on small plots of land (0.1 ha, or about the area over which shading and other competitive effects between trees might extend). Different tolerances of tree species to drought resulted in differential establishment, growth and survival of individuals of

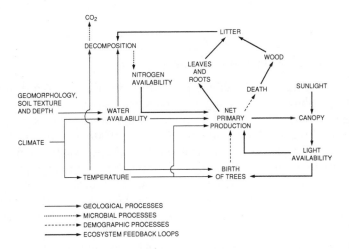

Figure 12.3 Schematic of model used by Pastor and Post (1988) to predict changes in northern forests exposed to climate change.

each species. The different species also had different rates of nitrogen uptake, nitrogen return to litter and release rates of nitrogen and carbon from litter.

The model was run for 200 years at 11 different sites during which the climate changed from the current climate to that for double the current amount of CO_2 and then run an additional 200 years with the new climate. In each case simulations were performed on sites that had sandy soils and sites that had silty clay loam soils. Initial soil carbon and nitrogen contents were set at 90% of those reported for boreal wet forest and cool temperate moist forest.

Major changes occurred on sites along the band of southern boreal and northern hardwood forests from the western Great Lakes eastward to the Atlantic Ocean. In general, on soils where there was no decrease in soil water availability with the doubling of CO_2, the current mixed spruce–fir–northern hardwood forest was replaced by a more productive northern hardwood forest. These northern hardwoods have a faster intrinsic growth rate and can attain greater steady-state biomass storage level than either spruce or fir. Also, the warmer climate as well as the higher nitrogen and lower lignin contents of northern hardwood litter enhanced nitrogen soil availability, which amplified the effect of warming on productivity.

On sites where soil water availability declined, the model spruce–fir–northern hardwood forest was replaced by a stunted pine–oak forest with much lower biomass storage. Although the warmer temperatures increased the decomposition rate and nitrogen mineralization, the higher lignin and lower nitrogen contents of oak and, particularly, pine litter compensated for this and reduced the decomposition rate.

A previous study (Solomon, 1986) did not consider climate-induced changes in N availability as a resultant effect of species composition changes, and thus did not show the dramatic changes in carbon storage that could result.

Effects on an ocean coastal upwelling system

Ocean nutrient fluxes are highly important to the yields of marine fisheries. Coastal upwelling zones are highly productive fishing grounds. Although such areas comprise less than 1% of the world's oceans, they supply more than 50% of the total fish harvest (Hartline, 1980). The reason for this richness is that nutrients are replenished from deeper waters by upwelling currents generated by a combination of local winds and topography. These nutrients then sustain an intense food web with important commercial fish at the top. There are several important upwelling areas in the world, the most notable being off the coast of Peru, providing the basis of the Peruvian anchovy fishery.

Changes in global climate, if they occur, will certainly have some effect on local wind patterns. One prediction is that the enhancement of daytime heating inland will cause alongshore winds on tropical and subtropical coasts to intensify and produce stronger upwellings (Bakun, 1990). Although this would mean a higher nutrient input and would seem to suggest greater primary productivity, it is not clear that this will increase primary productivity. Winds that are too strong can mix phytoplankton deeply and stir up sediment, reducing the radiation reaching the plankton. The complexity of coastal upwelling food webs is not understood well enough to permit the assumption that an increase in primary productivity, if it occurs, will translate into increased production of commercial fish, as it may be channelled into fish species of little commercial interest (Lasker, 1989). Intense winds may also affect fish production by preventing 'maintenance of fine-scale concentrations of minute food organisms essential to larval survival (Bakun, 1990). Thus, the effects of a sustained change in climate will have unpredictable effects on coastal upwelling fisheries.

12.4 SUMMARY AND CONCLUSIONS

This chapter returned to the global scale to consider some implications of the continuing input of CO_2 to the atmosphere from fossil fuel burning. This input of CO_2 is leading to increased levels of atmospheric CO_2, though not as rapidly as expected from the current level of understanding of carbon fluxes between the atmosphere and the terrestrial and oceanic components of the biosphere. This has stimulated the use of models to help understand these fluxes and predict their effects on the global carbon cycle as a whole.

Two processes in the global carbon cycle where more research may aid in explaining observations of CO_2 are the hypothesized fertilization effect of CO_2 on terrestrial plants and the biological pump in the ocean. A model of atmosphere–terrestrial plant interactions indicates that a fertilization effect on plants could buffer the atmosphere from some of the fossil fuel CO_2 input. However, other limiting nutrients, the effects of a higher plant decomposition rate, or a higher rate of herbivory could partially offset this buffering effect.

A pelagic ocean system model described the biological pump mechanism that transports carbon to the deep ocean in the form of faecal pellets. The mechanism is sensitive to the herbivore component of the pelagic food web. It is also sensitive to the influx of limiting nutrients (particularly nitrate) from below the euphotic zone. Other nutrients, such as iron, may limit primary production, and thus diminish the effect of the biological pump in the Antarctic Ocean.

The possible effects of climatic change (resulting from the greenhouse effect) on ecosystems is a matter of much concern. Computer simulations indicate that the change to warmer, drier weather in southern parts of boreal forests could cause a change in forest type, but this change could be towards either a higher productivity or lower productivity forest, depending on water and nitrogen availability on the particular site. Coastal upwelling fisheries, the source of one-half of production for human fisheries, will be affected by climate changes in ways that are difficult at present to predict.

The above considerations only begin to scratch the surface of the challenges presented by the human alteration of the global carbon cycle, which is only one of many current threats to the environment. Ecological food webs and the nutrient cycles that sustain them, and which they in turn affect, on scales ranging from local to global, may change dramatically over the next few decades due to human impacts. To understand and to control these changes will require a much greater commitment to both research and conservation than now exists.

Appendix A

The full solution of Equations (2.19a,b), for the initial values $C(0) = C_0$ and $C_s(0) = C_{s0}$, is

$$C(t) = \left[\left(\frac{\lambda_1 + C_5}{\lambda_1 - \lambda_2}\right)e^{\lambda_1 t} + \left(\frac{\lambda_2 + C_5}{\lambda_2 - \lambda_1}\right)e^{\lambda_2 t}\right]C_0$$

$$+ \left[\left(\frac{C_3}{\lambda_1 - \lambda_2}\right)e^{\lambda_1 t} + \left(\frac{C_3}{\lambda_2 - \lambda_1}\right)e^{\lambda_2 t}\right]C_{s0}$$

$$+ \frac{C_1}{C_2 C_5 - C_3 C_4}\left\{\left[\left(\frac{\lambda_1 + C_5}{\lambda_1 - \lambda_2}\right)e^{\lambda_1 t} + \left(\frac{\lambda_2 + C_5}{\lambda_2 - \lambda_1}\right)e^{\lambda_2 t} - 1\right]C_5\right.$$

$$\left. + \left[\left(\frac{C_3 C_4}{\lambda_1 - \lambda_2}\right)e^{\lambda_1 t} + \left(\frac{C_3 C_4}{\lambda_2 - \lambda_1}\right)e^{\lambda_2 t}\right]\right\}$$

(A.1a)

$$C_s(t) = \left[\left(\frac{C_4}{\lambda_1 - \lambda_2}\right)e^{\lambda_1 t} + \left(\frac{C_4}{\lambda_2 - \lambda_1}\right)e^{\lambda_2 t}\right]C_0$$

$$+ \left[\left(\frac{\lambda_1 + C_2}{\lambda_1 - \lambda_2}\right)e^{\lambda_1 t} + \left(\frac{\lambda_2 + C_2}{\lambda_2 - \lambda_1}\right)e^{\lambda_2 t}\right]C_{s0}$$

$$+ \left(\frac{C_1}{C_2 C_5 - C_3 C_4}\right)\left\{\left[\left(\frac{1}{\lambda_1 - \lambda_2}\right)e^{\lambda_1 t} + \left(\frac{1}{\lambda_2 - \lambda_1}\right)e^{\lambda_2 t}\right]C_5\right.$$

$$\left. + \left[\left(\frac{\lambda_1 + C_2}{\lambda_1 - \lambda_2}\right)e^{\lambda_1 t} + \left(\frac{\lambda_2 + C_2}{\lambda_2 - \lambda_1}\right)e^{\lambda_2 t} - 1\right]C_4\right\}$$

(A.1b)

where

$C_1 = I_n/V$

$C_2 = (q + vA)/V$

$C_3 = Ak_s/V$

$C_4 = vA(1 - k_u)/V_s$

$C_5 = Ak_s/V_s$

$\lambda_1 + \lambda_2 = -(C_2 + C_5)$

$\lambda_1 \lambda_2 = (C_2 C_5 - C_3 C_4)$

$$\lambda_{1,2} = -\left(\frac{1}{2}\right)\left(\frac{q}{V} + \frac{vA}{V} + \frac{k_s A}{V_s}\right) \pm \left(\frac{1}{2}\right)\left[\left(\frac{q}{V} + \frac{vA}{V} + \frac{k_s A}{V_s}\right)^2\right.$$

$$\left. - \frac{4Ak_s}{V_s V}(q + vk_u A)\right]^{1/2}$$

(A.2)

Appendix B

The solutions of the set of linear cascade Equations (2.21) for ovement of solute down a series of lakes can be found by solving Equation (2.21a) for C_1, inserting the solution into Equation (2.21b), solving that equation, and so forth.

The solution of Equation (2.21a) is

$$C_1(t) = C_{1,0} e^{-(q_1/V_1)t} + (q_1 C_{I,1} + J_1)(1 - R_1)(1 - e^{-(q_1/V_1)t})/q_1 \qquad \text{(B.1)}$$

where C_{10} is the initial concentration in Lake 1.

Equation (2.21b) is now

$$V_2 \frac{dC_2}{dt} = q_1(1 - R_2)\left[C_{10} e^{-(q_1/V_1)t} + \frac{(q_1 C_{I,1} + J_1)(1 - R_1)}{q_1}(1 - e^{-(q_1/V_1)t}) \right]$$

$$+ (q_2 C_{I,2} + J_2)(1 - R_2) - (q_1 + q_2)C_2$$

The solution for C_2 is

$$C_2(t) = C_{2,0} e^{-(q_1+q_2)t/V_2} + [A/(q_1 + q_2)](1 - e^{-(q_1+q_2)t/V_2})$$

$$+ \frac{B e^{-(q_1/V_1)t}}{[(q_1 + q_2)/V_2 - q_1/V_1]V_2}(1 - e^{-(q_1+q_2)t/V_2}) \qquad \text{(B.2)}$$

where

$$A = q_1(1 - R_2)(q_1 C_{I,1} + J_1)(1 - R_1)/(V_2 q_1) + (q_2 C_{I,2} + J_2)(1 - R_2)/V_2$$
$$B = q_1(1 - R_2)[C_{10} - (q_1 C_{I,1} + J_1)(1 - R_1)/q_1]/V_2$$

Appendix C

To linearize Equations (3.8a,b,c), substitute $N = N^* + N'$, $X = X^* + X'$, and $D = D^* + D'$ (as in 3.22a,b,c). Since N^*, X^* and D^* are constants, $d(N^* + N')/dt = dN'/dt$, $d(X^* + X')/dt = dX'/dt$, $d(D^* + D')/dt = dD'/dt$.

Consider Equation (3.8b) in detail. This becomes

$$\frac{dX'}{dt} = \frac{r_1(N^* + N')(X^* + X')}{k_1 + N^* + N'} - (d_1 + e_1)(X^* + X') \qquad (C.1)$$

When the right-hand side of Equation (C.1) is expanded and the terms of zero and first order in N, X and D are kept, this becomes

$$\frac{dX'}{dt} = \frac{r_1 N^* X'}{(k_1 + N^*)} - (d_1 + e_1)X^* + \frac{r_1 N^* X'}{k_1 + N^*} + \frac{r_1 N^* X'}{k_1 + N^*}$$

$$- \frac{r_1 N^* X^* N'}{(k_1 + N^*)^2} - (d_1 + e_1)X' \qquad (C.2)$$

The first two terms on the right-hand side of Equation (C.2) are zero-order terms. They cancel each other out because $N^* = k_1(d_1 + e_1)/(r_1 - d_1 - e_1)$. Also, terms four and five on the right-hand side of Equation (C.2) can be combined. Then, the linearized version is

$$\frac{dX'}{dt} = \frac{r_1 X^* N'}{k_1 + N^*} + \frac{r_1 k_1 N^* X'}{(k_1 + N^*)^2} - (d_1 + e_1)X' \qquad (C.3)$$

The same procedure can be used to linearize Equations (3.22a,c).

Appendix D

The determinant (3.24) can be expanded as

$$\left(-r_n - \frac{\gamma r_1 k_1 X^*}{(k_1 + N^*)^2} - \lambda\right)(-\lambda)(-d_2 - e_2 - \lambda) + \gamma d_1 d_2 \frac{r_1 k_1 X^*}{(k_1 + N^*)^2}$$

$$- (d_2 + e_2 + \lambda)\left(\frac{r_1 k_1 X^*}{(k_1 + N^*)^2}\right)\left(\frac{\gamma r_1 N^*}{k_1 + N^*}\right) = 0$$

or, from a rearrangement of terms

$$- \lambda^3 - \left[r_n + \frac{\gamma r_1 k_1 X^*}{(k_1 + N^*)^2} + d_2 + e_2\right]\lambda^2 - \left[\left(r_n + \frac{\gamma r_1 k_1 X^*}{(k_1 + N^*)^2}\right)(d_2 + e_2)\right.$$

$$\left. + \left(\frac{r_1 k_1 X^*}{(k_1 + N^*)^2}\right)\left(\frac{\gamma r_1 N^*}{k_1 + N^*}\right)\right]\lambda + \gamma d_2 d_1 \frac{r_1 k_1 X^*}{(k_1 + N^*)^2}$$

$$- (d_2 + e_2)\left(\frac{r_1 k_1 X^*}{(k_1 + N^*)^2}\right)\left(\frac{\gamma r_1 N^*}{k_1 + N^*}\right) = 0$$

Equation (3.9a) can be used to combine the last two terms. The result is

$$\lambda^3 + \left(r_n + \frac{\gamma r_1 k_1 X^*}{(k_1 + N^*)^2} + d_2 + e_2\right)\overset{\lambda^2}{} + \left[\left(r_n + \left(\frac{\gamma r_1 k_1 X^*}{(k_1 + N^*)^2}\right)(d_2 + e_2)\right.\right.$$

$$\left.\left. + \left(\frac{r_1 k_1 X^*}{(k_1 + N^*)^2}\right)\left(\frac{\gamma r_1 N^*}{(k_1 + N^*)}\right)\right]\lambda + (e_1 d_2 + e_1 d_1 + e_1 e_2)\frac{\gamma r_1 k_1 X^*}{(k_1 + N^*)^2} = 0$$

All of the terms in this equation for λ are positive. A theorem of algebra states that a necessary and sufficient condition for the real parts of all of the eigenvalues of an nth order equation of the form above (with real coefficients) to be negative is that the coefficients all be positive (e.g. Murata, 1977; see Chapter 3, Theorem 9).

References

Aandahl, A. R. (1949) The characteristics of slope positions and their influence on the total nitrogen content of a few virgin soils of western Iowa. *Proc. Soil Sci. Soc. Am.*, **13**, 449–54.

Aber, J. D. and Melillo, J. M. (1979) Litter decomposition: Measuring relative contributions of organic matter and nitrogen to forest soils. *Can. J. Bot.*, **58**, 416–21.

Aber, J. D. and Melillo, J. M. (1982) Nitrogen immobilization in decaying hardwood leaf litter as a function of initial nitrogen and lignin content. *Can. J. Bot.* **60**, 2263–9.

Abrams, P. A. (1988) How should resources be counted? *Theor. Popul. Biol.*, **33**, 226–42.

Adams, S. M., Kimmel, B. L. and Ploskey, G. R. (1983) Sources of organic matter for reservoir production: A trophic-dynamics analysis. *Can. J. Fish. Aquat. Sci.*, **40**, 1480–95.

Ågren, G. I. (1985) Limits to plant production. *J. Theor. Biol.*, **113**, 89–92.

Ågren, G. I. and Axelsson, B. (1980) Population respiration: a theoretical approach. *Ecol. Modell.*, **11**, 39–55.

Ahlgren, I. (1980) A dilution model applied to a system of shallow eutrophic lakes after diversion of sewage effluents. *Arch. Hydrobiol.*, **89**, 17–32.

Allen, T. F. H. and Starr, T. B. (1982) *Hierarchy: Perspective for Ecological Complexity.* University of Chicago Press, Chicago, 310 pp.

Andersen, P. and Fenchel, T. (1985) Bactivory by microheterotrophic flagellates in seawater samples. *Limnol. Oceanogr.*, **30**, 186–202.

Andersen, P. and Soerensen H. M. (1986) Population dynamics and trophic coupling in pelagic microorganisms in eutrophic coastal waters. *Mar. Ecol. Progr. Ser.*, **33**. 99–109.

Anderson, D. H. (1983) *Compartment Modeling and Tracer Kinetics. Lecture Notes in Biomathematics 50.* Springer-Verlag, Berlin, West Germany. 302 pp.

Anderson, F. O. (1978) Effects of nutrient level on the decomposition of *Phragmites communis* Trin. *Arch. Hydrobiol.*, **84**, 42–54.

Anderson, J. M., Ineson, P. and Huish, S. A. (1983) Nitrogen and cation mobilization by soil fauna on leaf litter and soil organic matter from deciduous woodlands. *Soil Biol. Biochem.*, **15**, 463–67.

Andrewartha, H. G. and Birch, L. C. (1984) *The Distribution and Abundance of Animals. The Ecological Web.* University of Chicago Press, Chicago.

Armstrong, R. A. (1976) Fugitive species: Experiments with fungi and some theoretical considerations. *Ecology*, **57**, 953–63.

Armstrong, R. A. (1979) Prey species replacement along a gradient of nutrient enrichment: A graphical approach. *Ecology*, **60**, 76–84.

Armstrong, R. A. and McGehee, R. (1980) Competitive exclusion. *Am. Nat.*, **115**, 151–70.

Attiwill, P. M. (1980) Nutrient cycling in a *Eucalyptus obliqua* (L'Herit.) forest. IV. Nutrient uptake and nutrient return. *Austr. J. Bot.*, **28**, 199–222.

Auer, M. T., Kieser, M. S. and Canale, R. P. (1986) Identification of critical nutrient levels through field verification of models for phosphorus and phytoplankton growth. *Can. J. Fish. Aquat. Sci.*, **43**, 379–88.

Austin, M. P. and Cook, B. G. (1974) Ecosystem stability: A result from an abstract simulation. *J. Theor. Biol.*, **45**, 435–58.

Bacastow, R. and Keeling, C. D. (1981) Atmospheric carbon dioxide and radiocarbon in the natural carbon cycle: II. Changes from A.D. 1700 to 2070 as deduced from a geochemical model. In *Carbon and the Biosphere* (eds. G. M. Woodwell and E. V. Pecan), pp. 86–135. U.S. Atomic Energy Commission, Washington, D.C.

Bader, F. G., Frederickson, A. G. and Tsuchiya, H. M. (1976) Dynamics of an algal-protozoan grazing interaction. In *Modeling Biochemical Processes in Aquatic Ecosystems* (ed. R. P. Canale), pp. 257–280. Ann Arbor Science, Ann Arbor, Michigan.

Bakelaar, R. G. and Odum, E. P. (1978) Community and population level responses to fertilization in an old-field ecosystem. *Ecology*, **59**, 660–5.

Bakun, A. (1990) Global climate change and intensification of coastal ocean upwelling. *Science*, **247**, 198–201.

Bailey, N. T. J. (1964) *The Elements of Stochastic Processes: With Applications to the Natural Sciences*. John Wiley and Sons, New York, 249 pp.

Ballard, L. A. T. (1941) The depressant effect of carbon dioxide on photosynthesis. *New Phytol.*, **40**, 276–90.

Barkley, S. A., Barel, D., Stoner, W. A. and Miller, P. C. (1978) Controls on decomposition and mineral release in wet meadow tundra – a simulation model in Environmental Chemistry and Cycling Processes: Proceedings of Symposium, Augusta, Georgia.

Barlow, J. P. and Bishop, J. W. (1965) Phosphate regeneration by zooplankton in Cayuga Lake. *Limnol. Oceanogr.*, **10** (suppl.), R15–R24.

Barsdate, R. J., Prentki, R. T. and Fenchel, T. (1974) Phosphorus cycle of model ecosystems: Significance for decomposer food chains and effect of bacteria grazers. *Oikos*, **25**, 239–51.

Bartell, S. M. (1981) Potential impact of size-selective planktivory on phosphorus release by zooplankton. *Hydrobiologia*, **80**, 139–45.

Bartell, S. M. and Kitchell, J. F. (1978) Seasonal impact of planktivory on phosphorus release by Lake Wingra zooplankton. *Verh. Int. Verein. Limnol.*, **20**, 466–74.

Bartholomew, W. V. (1965) Mineralization and immobilization of nitrogen in the decomposition of plant and animal residues. In *Soil Nitrogen. Agronomy 10* (eds. W. V. Bartholomew and F. E. Clark), pp. 285–306. American Society of Agronomy, Madison, Wisconsin.

Batzli, G. O., White, R. G., MacLean, Jr., S. F., Pitelka, F. A. and Collier, B. D. (1980) The herbivore-based trophic system. In *The Coastal Tundra at Barrow, Alaska* (eds. J. Brown, P. C. Miller, L. L. Tieszen and F. L. Bunnell), pp. 335–410. US/IBP Synthesis Series 12. Dowden, Hutchinson & Ross, Stroudsburg, Pennsylvania.

Beadle, N. C. W. (1954) Soil phosphate and the delimitation of plant communities in eastern Australia. *Ecology*, **35**, 370–75.

Begon, M. and Mortimer, M. (1981) *Population Ecology: A Unified Study of Animals and Plants*. Sinauer Associates, Sunderland, Massachusetts. 200 pp.

Belovsky, G. E. and Jordan, P. A. (1981) Sodium dynamics and adaptations of a moose population. *J. Mammol.*, **62**, 613–21.

Bender, E. A., Case, T. J. and Gilpin, M. E. (1984) Perturbation experiments in community ecology: Theory and practice. *Ecology*, **65**, 1–13.

Berendse, F. (1985) The effect of grazing on the outcome of competition between plant species with different nutrient requirements. *Oikos*, **44**, 35–9.

Berg, B. and Ekbohm, G. (1983) Nitrogen immobilization in decomposing needle litter at variable carbon:nitrogen ratios. *Ecology*, **64**, 63–7.

Berg, H. C. (1983) *Random Walks in Biology*. Princeton University Press, Princeton, 142 pp.

Berner, R. A. and Lasaga, A. C. (1989) Modeling the geochemical carbon cycle. *Sci. Am.*, **260(3)**, 74–81.

Bierman, V. J., Jr. (1976) Mathematical model of the selective enhancement of blue–green algae by nutrient enrichment. In *Modeling Biochemical Processes in Aquatic Ecosystems* (ed. R. P. Canale), pp. 1–32. Ann Arbor Science Publishers, Ann Arbor, Michigan.

Bjornsen, P. K., Riemann, B., Horsted, S. J., Nielsen, T. G. and Pock-Sten, J. (1988) Trophic interactions between heterotrophic nanoflagellates and bacterioplankton in manipulated seawater enclosures. *Limnol. Oceanogr.*, **33**, 409–20.

Black, A. L. and Wight, J. R. (1979) Range fertilization: Nitrogen and phosphorus uptake and recovery over time. *J. Range Manag.*, **32**, 349–53.

Bliss, L. C. and Wein, R. W. (1972) Plant community responses to disturbances on the western Canadian Arctic. *Can. J. Bot.*, **50**, 1097–109.

Bolla, M. and Kutas, T. (1984) Submodels for the nutrient loading estimation on River Zala. *Ecol. Modell.*, **26**, 115–43.

Boraas, M. E. (1980) A chemostat system for the study of rotifer–algal–nitrate interactions. In *Evolution and Ecology of Zooplankton Communities*. (ed. W. C. Kerfoot), pp. 173–82. University Press of New England, Hanover, New Hampshire.

Borgmann, U., Millard, E. S. and Charlton, C. C. (1988) Dynamics of a stable, large volume, laboratory ecosystem containing *Daphnia* and phytoplankton. *J. Plankton Res.*, **10**, 691–713.

Boring, L. R., Monk, C. D. and Swank, W. T. (1981) Early regeneration of a clearcut Appalachian forest. *Ecology*, **62**, 1244–53.

Bormann, F. H. and Likens, G. E. (1979) *Pattern and Process in a Forested Ecosystem*. Springer-Verlag, New York.

Bosatta, E. and Staaf, H. (1982) The control of nitrogen turnover in forest litter. *Oikos*, **39**, 143–51.

Bosman, A. L., Hockey, P. A. R. and Siegfried, W. R. (1987) The influence of coastal upwelling on the functional structure of rocky intertidal communities. *Oecologia (Berlin)*, **72**, 226–32.

Botkin, D. B. (1990) *Discordant Harmonies*. Oxford University Press, Oxford, 241 pp.

Botkin, D. B. and Sobel, M. J. (1975) Stability in time-varying ecosystems. *Am. Nat.*, **109**, 625–46.

Boyd, C. E. and Lawrence, J. M. (1966) The mineral composition of several freshwater algae. *Proc. Annu. Conf. Southeastern Assoc. Game Fish.*, **20**, 413–24.

Boynton, W. R., Kemp, W. M. and Keefe, C. W. (1982) A comparative analysis of nutrients and other factors influencing estuarine phytoplankton production. In *Estuarine Comparisons*, pp. 69–90. Academic Press, New York.

Braakhekke, W. G. (1980) On coexistence: A causal approach to diversity and stability in grassland vegetation. *Agricultural Research Reports* 902, Centre for Agrobiological Research, Wageningen, The Netherlands. 164 pp.

Broadbent, F. E. and Nakashima, T. (1971) Effect of added salts on nitrogen mineralization in three California soils. *Proc. Soil Sci. Soc. Am.*, **35**, 457–60.

Broecker, W. S., Peng, T.-H. and Ostlund, G. (1986) The distribution of bomb radiocarbon in the ocean. *J. Geophys. Res.*, **91**, 14331–44.

Brooks, J. L. and Dodson, S. I. (1965) Predation, body size, and composition of plankton. *Science*, **150**, 28–35.

Brown, P. L. and Dickey, D. D. (1970) Losses of wheat straw residue under simulated field conditions. *Proc. Soil Sci. Soc. Am.*, **34**, 118–21.

Brunsting, A. M. H. and Heil, G. W. (1985) The role of nutrients in the interactions between a herbivorous beetle and some competing plant species in heathlands. *Oikos*, **44**, 23–6.

Buechler, D. G. and Dillon, R. D. (1974) Phosphorus regeneration in fresh-water paramecia. *J. Protozool.*, **21**. 339–43.

Burmaster, D. E. (1979) The continuous culture of phytoplankton: Mathematical equivalence among three steady-state models. *Amer. Nat.*, **113**, 123–34.

Button, D. K. (1978) On the theory of control of microbial growth kinetics by limiting nutrient concentrations. *Deep-Sea Res.*, **25**, 1163–77.

Canfield, D. E., Green, W. J., Gardner, T. J. and Ferdelman, T. (1984) Elemental residence times in Acton Lake, Ohio. *Arch. Hydrobiol.*, **100**, 501–19.

Caperon, J. W. (1965) The dynamics of nitrate limited growth of *Isochrysis galbana* populations. Ph.D. Thesis, Scripps Institution of Oceanography, La Jolla, California. 71 pp.

Caperon, J. W. (1967) Population growth in micro-organisms limited by food supply. *Ecology*, **48**, 715–22.

Caperon, J. (1968) Population growth response of *Isochrysis galbana* to variable nitrate environment. *Ecology*, **49**, 866–72.

Caperon, J. W. and Meyer, J. (1972a) Nitrogen-limited growth of marine phytoplankton. I. Changes in population characteristics with steady-state growth rate. *Deep-Sea Res.*, **19**, 601–18.

Caperon, J. W. and Meyer, J. (1972b) Nitrogen-limited growth of marine phytoplankton. II. Uptake kinetics and their role in nutrient-limited growth of phytoplankton. *Deep-Sea Res.*, **19**, 619–32.

Capinera, J. L. and Roltsch, W. J. (1980) Response of wheat seedlings to actual and simulated migratory grasshopper defoliation. *J. Econ. Entomol.*, **73**, 258–61.

Caraco, N., Tamse, A., Boutros, O. and Valiela, I. (1987) Nutrient limitation of phytoplankton growth in brackish coastal ponds. *Can. J. Fish. Aquat. Sci.*, **44**, 474–6.

Carpenter, S. R. (1980) Enrichment of Lake Wingra, Wisconsin, by submersed macrophyte decay. *Ecology*, **61**, 1145–55.

Carpenter, S. R. (1981) Decay of heterogeneous detritus: a general model. *J. Theor. Biol.*, **89**, 539–47.

Carpenter, S. R. and Kitchell, J. F. (1984) Plankton community structure and limnetic primary production. *Am. Nat.*, **124**, 159–72.

Carpenter, S. R., Kitchell, J. F. and Hodgson, J. R. (1985) Cascading trophic interactions and lake productivity. *BioScience*, **35**, 634–9.

Caughley, G. (1976) Plant–herbivore systems. In *Theoretical Ecology: Principles and Applications* (ed. R. M. May), pp. 94–113. W. B. Saunders Company, Philadelphia, Pennsylvania.

Challinor, J. L. and Gersper, P. L. (1975) Vehicle perturbation effects upon a tundra soil–plant system. II. Effects on the chemical regime. *Proc. Soil Sci. Soc. Am.*, **39**, 689–95.

Chapin, F. S., III, Barsdate, R. J. and Barel, D. (1978) Phosphorus cycling in Alaskan coastal tundra: A hypothesis for the regulation of nutrient cycling. *Oikos*, **31**, 189–99.

Chapin, F. S., III, Johnson, D. A. and McKendrick, J. D. (1980) Seasonal movement of nutrients in plants of differing growth form in an Alaskan tundra ecosystem: Implications for herbivory. *J. Ecol.*, **68**, 189–209.

Chapin, F. S., III and Van Cleve, K. (1978) Nitrogen and phosphorus distribution in an Alaskan tussock tundra ecosystem: natural patterns and implications for development. In *Environmental Chemistry and Cycling Processes: Proceedings of Symposium, Augusta, Georgia, April 28–May 1, 1976*. (eds D. C. Adriano and I. L. Brisbin), pp. 738–753. U.S. Department of Energy, CONF-760429.

Chapra, S. C. (1975) Comment on 'An empirical method of estimating retention of phosphorus in lakes' by W. B. Kirchner and P. J. Dillon. *Water Resour. Res.*, **11**, 1033–4.

Chesson, P. L. and Warner, R. R. (1981) Environmental variability promotes coexistence in lottery competitive systems. *Am. Nat.*, **117**, 923–43.

Child, G. I. and Shugart, Jr., H. H. (1972) Frequency response analysis of magnesium cycling in a tropical forest ecosystem. In *Systems Analysis and Simulation in Ecology* (ed. B. C. Patten), Vol. II. Academic Press, New York.

Clarkson, D. T. (1967) Phosphorus supply and growth rate in species of *Agrostis* L. *J. Ecol.*, **55**, 111–18.

Cole, C. V., Elliott, E. T., Hunt, H. W. and Coleman, D. C. (1978) Trophic interactions in soil as they affect energy and nutrient dynamics. V. Phosphorus transformations. *Microb. Ecol.*, **4**, 381–7.

Coley, P. D., Bryant, J. P. and Chapin, III, F. S. (1985) Resource availability and plant antiherbivore defense. *Science*, **230**, 895–9.

Colinvaux, P. (1986) *Ecology*. John Wiley and Sons, New York. 725 pp.

Collins, C. D. (1980) Formulation and validation of a mathematical model of phytoplankton growth. *Ecology*, **61**, 639–49.

Connell, J. H. (1961) The influence of interspecific competition and other factors on the distribution of the barnacle *Chthalamus stellatus*. *Ecology*, **42**, 710–23.

Connell, J. H. (1978) Diversity in tropical rainforests and coral reefs. *Science*, **99**, 1302–10.

Cooper, D. C. (1973) Enhancement of net primary productivity by herbivore grazing in aquatic laboratory microcosms. *Limnol. Oceanogr.*, **18**, 31–7.

Coppock, D. L., Detling, J. K., Ellis, J. E. and Dyer, M. I. (1983a) Plant–herbivore interactions in a North American mixed-grass prairie. I. Effects of black-tailed prairie dogs on intra-seasonal aboveground plant biomass and nutrient dynamics and plant species diversity. *Oecologia*, **56**, 1–9.

Coppock, D. L., Detling, J. K., Ellis, J. E. and Dyer, M. I. (1983b) Plant–herbivore interactions in a North American mixed-grass prairie. II. Responses of bison to modifications of vegetation by prairie dogs. *Oecologia*, **56**, 10–15.

Crank, J. (1975) *The Mathematics of Diffusion*, 2nd edn, Oxford University Press, Oxford, 414 pp.

Crawley, M. J. (1973) The numerical responses of insect predators to changes in prey density. *J. Anim. Ecol.*, **44**, 877–92.

Crawley, M. J. (1983) *Herbivory: The Dynamics of Animal–Plant Interactions*. University of California Press, Berkeley, California, 437 pp.

Crombie, A. C. (1947) Interspecific competition. *J. Anim. Ecol.*, **16**, 44–73.

Csanady, G. T. (1973) *Turbulent Diffusion in the Environment*. D. Reidel, Boston.

Cuker, B. E. (1983) Grazing and nutrient interactions in controlling the activity and composition of the epilithic algal community of an arctic lake. *Limnol. Oceanogr.*, **28**, 133–41.

Cushing, D. H. (1959) On the nature of production in the sea. *Fish. Invest. (Lond.) Ser. II*, **18(7)**. 104 pp.

Cushing, D. H. (1961) On the failure of the Plymouth herring fishery. *J. Mar. Biol. Assoc. U.K.*, **41**, 799–816.

Cushing, D. H. (1968) Grazing by herbivorous copepods in the sea. *J. Cons. Int. Explor. Mer.*, **32**, 70–82.

Danell, K., Huss-Danell, K. and Bergstrom, R. (1985) Interactions between browsing moose and two species of birch in Sweden. *Ecology*, **66**, 1867–78.

Davis, S. M. (1989) Sawgrass and cattail production in relation to nutrient supply in the Everglades. In *Freshwater Wetlands and Wildlife* (eds R. R. Sharitz and J. W. Gibbons), pp. 325–341. Office of Scientific and Technical Information, U.S. Department of Energy, Oak Ridge, Tennessee.

Davy, A. J. and Bishop, G. F. (1984) Response of *Hieracium pilosella* in Breckland grass-heath to inorganic nutrients. *J. Ecol.*, **72**, 319–30.

Day, J. W., Hall, C. A. S., Kemp, W. M. and Yanez-Arancibia, A. (1989) *Estuarine Ecology*. John Wiley and Sons, New York, 558 pp.

DeAngelis, D. L. (1975) Stability and connectance in food web models. *Ecology*, **56**, 238–43.

DeAngelis, D. L. (1980) Energy flow, nutrient cycling, and ecosystem resilience. *Ecology*, **61**, 764–71.

DeAngelis, D. L., Bartell, S. M. and Brenkert, A. L. (1989) Effects of nutrient recycling and food-chain length on resilience. *Nat.*, **134**, 778–805.

DeAngelis, D. L., Goldstein, R. A. and O'Neill, R. V. (1975) A model for trophic interaction. *Ecology*, **56**, 881–92.

DeAngelis, D. L., Post, W. M. and Travis, C. C. (1986) *Positive Feedback in Natural Systems*. Springer-Verlag, Berlin, 290 pp.

DeAngelis, D. L. and Waterhouse, J. C. (1987) Equilibrium and non equilibrium concepts in ecological models. *Ecological monographs*, **57**, 1–21.

DeAngelis, D. L., Waterhouse, J. C., Post, W. M. and O'Neill, R. V. (1985) Ecological modelling and disturbance evaluation. *Ecol. Modell.*, **29**, 339–419.

D'Elia, C. F. (1988) The cycling of essential elements in coral reefs. In *Concepts of Ecosystem Ecology: A Comparative View* (eds L. R. Pomeroy and J. J. Alberts), pp. 195–230. Springer–Verlag, New York.

DeMott, W. R. (1982) Feeding selectivities and relative ingestion rates of *Daphnia* and *Bosmina*. *Limnol. Oceanogr.*, **27**, 518–27.

Denn, M. M. (1986) *Process Modeling*. Pitman Publishing, Marshfield, Massachusetts, 324 pp.

Denny, M. W. (1988) *Biology and the Mechanics of the Wave-Swept Environment*. Princeton University Press, Princeton, New Jersey. 329 pp.

deNoyelles, F., Jr. and O'Brien, W. J. (1978) Phytoplankton succession in nutrient enriched experimental ponds as related to changing carbon, nitrogen and phosphorus conditions. *Arch. Hydrobiol.*, **84**, 137–65.

DePinto, J. V., Young, T. C., Bonner, J. S. and Rodgers, P. W. (1986) Microbial recycling of phytoplankton phosphorus. *Can. J. Fish. Aquat. Sci.*, **43**, 336–42.

Detling, J. K., Dyer, M. I. and Winn, D. T. (1979) Effect of simulated grasshopper grazing on CO_2 exchange rates of western wheatgrass leaves. *J. Econ. Entomol.*, **72**, 403–6.

Detling, J. K. and Dyer, M. I. (1981) Evidence for potential plant growth regulator in grasshoppers. *Ecology*, **62**, 485–8.

Detwiler, R. P., Hall, C. A. S. and Bogdnoff, P. (1985) Land use change and carbon exchange in the tropics: II. Estimates for the entire region. *Environ. Manag.*, **9**, 335–44.

DiToro, D. M. (1976) Combining chemical equilibrium and phytoplankton models – a general methodology. In *Modeling Biochemical Processes in Aquatic Ecosystems* (ed. R. P. Canale), pp. 233–256. Ann Arbor Science Publishers, Ann Arbor, Michigan.

DiToro, D. M. (1980) Applicability of cellular equilibrium and Monod theory to phytoplankton growth kinetics. *Ecol. Model.*, **8**, 201–18.

DiToro, D. M., O'Connor, D. J. and Mancini, J. L. (1971) A dynamic model of the phytoplankton population in the Sacramento–San Joaquin delta. *Adv. Chem. Ser.*, **106**, 131–80.

DiToro, D. M., Thomann, R. V., O'Connor, D. J. and Mancini, J. L. (1977) Estuarine phytoplankton biomass models – verification analysis and preliminary applications. In *The Sea*, (eds E. Goldberg, J. Steele and J. J. O'Brien) Vol. 6, pp. 969–1020. Wiley, New York.

Dixon, K. R., Luxmoore, R. J. and Begovich, C. L. (1978) CERES: A model of forest stand biomass dynamics for predicting tree contaminant, nutrient, and water effects. ORNL/EATC-25, Oak Ridge National Laboratory, Oak Ridge, Tennessee. 102 pp.

Donaldson, J. R. (1967) Phosphorus budget of Iliamna Lake, Alaska, as related to the cyclic abundance of sockeye salmon. Ph.D. Dissertation, University of Washington.

Douce, G. K. and Webb, D. P. (1978) Indirect effects of soil invertebrates on litter decomposition: Elaboration via analysis of a tundra model. *Ecol. Modell.*, **4**, 339–59.

Drent, R. (1980) Goose flocks and food exploitation: How to have your cake and eat it. In *Acta XVII Congressus Internationalis Ornithologici* (ed. R. Nohring), Vol. II, pp. 800–806. Deutschen Ornithologen-Gesellschaft, Berlin.

Droop, M. R. (1968) Vitamin B_{12} and marine ecology. IV. The kinetics of uptake, growth and inhibition in *Monochrysis lutheri*. *J. Mar. Biol. Assoc. U.K.*, **48**, 689–733.

Droop, M. R. (1973) Some thoughts on nutrient limitation in algae. *J. Phycol.*, **9**, 264–72.

Droop, M. R. (1974) The nutrient status of algal cells in a continuous culture. *J. Mar. Biol. Assoc. U.K.*, **54** 825–55.

Drury, W. H. and Nisbet, C. T. (1973) Succession. *J. Arnold Arboretum*, **54**, 331–68.

Dudzik, M., Harte, J., Levy, D. and Sandusky, J. (1975) Stability indicators for nutrient cycles in ecosystems. LBL-3265. Berkeley, California: Lawrence Berkeley Laboratory, University of California.

Dugdale, R. C. (1967) Nutrient limitation in the sea: Dynamics, identification, and significance. *Limnol. Oceanogr.*, **12**, 685–95.

Dugdale, R. C. and Goering, J. J. (1967) Uptake of new and regenerated forms of nitrogen in primary productivity. *Limnol. Oceanogr.*, **12**, 196–206.

Dyer, M. I. (1975) The effects of red-winged blackbirds (*Agelaius phoeniceus* L.) on biomass production of corn (*Zea mays* L.). *J. Appl. Ecol.*, **12**, 719–26.

Dyer, M. I., DeAngelis, D. L., and Post, W. M. (1986) A model of herbivore feedback on plant productivity. *Mathemat. Biosci.*, **79**, 171–84.

Dyer, M. I., Detling, J. K., Coleman, D. C. and Hilbert, D. W. (1982) The role of herbivores in grasslands. In *Grasses and Grasslands* (eds J. R. Estes, R. J. Tyrl and J. N. Brunken) pp. 255–96. University of Oklahoma Press, Norman, Oklahoma.

Dykyjova, D. and Kvet, J. (1978) *Pond Littoral Ecosystems*. Springer-Verlag, New York.

Edmondson, W. T. and Litt, A. H. (1982) *Daphnia* in Lake Washington. *Limnol. Oceanogr.*, **27**, 272–93.

Elliott, E. T., Castanares, L. G., Perlmutter, D. and Porter, K. G. (1983) Trophic-level control of production and nutrient dynamics in an experimental planktonic community. *Oikos*, **41**, 7–16.

Elton, C. S. (1927) *Animal Ecology*. Macmillan, New York.

Elwood, J. W., Newbold, J. D., O'Neill, R. V. and Van Winkle, W. (1983) Resource spiraling: an operational paradigm for analyzing lotic ecosystems. pp. 3–27. In *Dynamics of Lotic Ecosystems* (eds T. D. Fontaine, III and S. M. Bartell), pp. 3–27. Ann Arbor Science, Ann Arbor, Michigan.

Eppley, R. W. and Thomas, W. H. (1969) Comparison of half-saturation 'constants' for growth and nitrate uptake of marine phytoplankton. *J. Phycol.*, **5**, 365–9.

Eppley, R. W., Venrick, E. L. and Mullin, M. M. (1973) A study of plankton dynamics and nutrient cycling in the central gyre of the N. Pacific ocean. *Limnol. Oceanogr.*, **18**, 534–51.

Ericksson, E. and Welander, P. (1956) On a mathematical model of the carbon cycle in nature. *Tellus*, **8**, 155–75.

Fenchel, T. (1982) Ecology of heterotrophic microflagellates. II. Bioenergetics of growth. *Mar. Ecol. Prog. Ser.*, **8**, 225–31.

Fife, D. N. and Nambiar, E. K. S. (1982) Accumulated retranslocation of mineral nutrients in developing needles in relation to seasonal growth in young radiata pine trees. *Ann Bot.*, **50**, 817–29.

Finn, J. T. (1976) Measures of ecosystem structure and function derived from analysis of flows. *J. Theor. Biol.*, **56**, 363–80.

Finn, J. T. (1978) Cycling index: a general definition for cycling in compartment models. In *Environmental Chemistry of and the Cycling Process* (eds D. C. Adriano and I. L. Brisbin), pp. 138–64. DOE Symposium Series 45, CONF-760429. National Technical Information Sevice, Springfield, Virginia.

Finn, J. T. (1980) Flow analysis of models of the Hubbard Brook Ecosystem. *Ecology*, **61**, 572–9.

Finn, J. T. (1982) Ecosystem succession, nutrient cycling and output-input ratios. *J. Theor. Biol.*, **99**, 479–89.

Fox, J. F. and Bryant, J. P. (1984) Instability of the snowshoe hare and wood plant interaction. *Oecologia (Berlin)*, **63**, 128–35.

Fretwell, S. D. (1977) The regulation of plant communities by the food chains exploiting them. *Perspect. Biol. Med.*, **20**, 169–85.

Fretwell, S. D. (1987) Food chain dynamics: The central theory of ecology? *Oikos*, **50**, 291–301.

Fuhs, G. W. (1969) Phosphorus content and rate of growth in the diatoms *Cyclotella nana* and *Thalassiorira fluviatilis*. *J. Phycol.*, **5**, 312–21.

Funderlic, R. E. and Heath, M. T. (1971) Linear compartmental analysis of ecosystems. ORNL-IBP-71-4. International Biological Program. Oak Ridge National Laboratory, Oak Ridge, Tennessee.

Gallepp, G. W. (1979) Chironomid influence on phosphorus release in sediment-water microcosms. *Ecology*, **60**, 547–56.

Gallopin, G. C. (1971a) A generalized model of a resource–population system. I. General properties. *Oecologia (Berlin)*, **7**, 382–413.

Gallopin, G. C. (1971b) A generalized model of a resource–population system. II. Stability analysis. *Oecologia (Berlin)*, **7**, 414–32.

Gates, C. T. and Wilson, J. R. (1974) The interaction of nitrogen and phosphorus on the growth, nutrient status, and modulation of *Stylosanthes humilis* H. B. K. (Townsville Stylo). *Plant Soil*, **41**, 325–33.

Gatto, M. and Rinaldi, S. (1987) Some models of catastrophic behavior in exploited forests. *Vegetatio*, **69**, 213–22.

Gause, G. F. (1934) *The Struggle for Existence*. Williams and Wilkins, Baltimore. (Reprinted 1965, by Hafner, New York.)

Gholz, H. L., Perry, C. S., Cropper, Jr., W. P. and Hendry, L. C. (1985) Litterfall, decomposition, and nitrogen and phosphorus dynamics in a chronosequence of slash pine (*Pinus elliotii*) plantations. *For. Sci.*, **31**, 463–78.

Gilpin, M. E. (1972) Enriched predator–prey systems: theoretical stability. *Science*, **177**, 902–4.

Gilpin, M. E. (1975) *Group Selection in Predator-Prey Communities*. Princeton University Press, Princeton, New Jersey.

Gliwicz, Z. M. (1977) Food size selection and seasonal succession of filter feeding zooplankton in an eutrophic lake. *Ekol. Pol.*, **25**, 179–225.

Gliwicz, Z. M. (1980) Filtering rates, food size selection, and feeding rates in cladocerans – another aspect of interspecific competition in filter-feeding zooplankton. In *Evolution and Ecology of Zooplankton Communities* (ed. W. C. Kerfoot), pp. 282–291. University Press of New England, Hanover, New Hampshire.

Golkin, K. R. and Ewel, K. C. (1984) A computer simulation of the carbon, phosphorus, and hydrological cycles of a pine flatwoods ecosystem. *Ecol. Modell.*, **24**, 113–36.

Golterman, H. L., Bakels, C. C. and Jacobs-Mogelin, J. (1969) Availability of mud phosphates for the growth of algae. *Verh. Int. Ver. Limnol.*, **19**, 182–8.

Goodman, G. T. and Perkins, D. F. (1959) Mineral uptake and retention in cotton-grass (*Eriophorum vaginatum* L.). *Nature*, **184**, 467–8.

Gordon, A. M. and Van Cleve, K. (1983) Seasonal patterns of nitrogen mineralization following harvesting in the white spruce forests of interior Alaska. In *Resources and Dynamics of the Boreal Zone* (eds. R. W. Wein, R. R. Pierce and I. R. Methven) pp. 119–130. Association of Canadian Universities for Northern Studies, Sault Sainte Marie, Ontario.

Gorham, E., Vitousek, P. M. and Reiners, W. A. (1979) The regulation of chemical budgets over the course of terrestrial ecosystem succession. *Annu. Rev. Ecol. System.*, **10**, 53–84.

Gosz, J. R. (1981) Nitrogen cycling in coniferous ecosystems. In *Nitrogen Cycling in Terrestrial Ecosystems: Processes, Ecosystem Strategies, and Management Implications* (eds F. E. Clark and T. H. Rosswall), pp. 405–426. Ecological Bulletin 33, Swedish Natural Science Research Council, Stockholm.

Gray, J. T. (1983) Nutrient use by evergreen and deciduous shrubs in Southern California. I. Community nutrient cycling and nutrient use efficiency. *J. Ecol.*, **71**, 21–41.

Gray, J. T. and Schlesinger, W. H. (1983) Nutrient use by evergreen and deciduous shrubs in Southern California. II. Experimental investigations of the relationship between growth, nitrogen uptake and nitrogen availability. *J. Ecol.*, **71**, 43–56.

Grenny, W. J., Bella, D. A. and Curl, Jr., H. C. (1974) Effects of intracellular nutrient pools on growth dynamics of phytoplankton. *Water Pollut. Contr. Fed.*, **46**, 1751–60.

Grubb, P. J. (1977) The maintenance of species richness in plant communities: the importance of the regeneration niche. *Biol. Rev.*, **52**, 107–145.

Gutschick, V. P. (1981) Evolved strategies in nitrogen acquisition by plants. *Am. Nat.*, **118**, 607–37.

Hairston, N. G., Smith, F. E. and Slobodkin, L. B. (1960) Community structure, population control and competition. *Am. Nat.*, **94**, 421–5.

Hall, C. A. S. (1972) Migration and metabolism in a temperate stream ecosystem. *Ecology*, **53**, 585–604.

Hall, D. J., Threlkeld, S. T., Burns, C. W. and Crowley, P. H. (1976) The size-efficiency hypothesis and the size structure of zooplankton communities. *Annu. Rev. Ecol. System.*, **7**, 177–203.

Hallam, T. G. (1978) Structural sensitivity of grazing formulation in nutrient controlled plankton models. *J. Mathemat. Biol.*, **5**, 269–80.

Harmsen, G. W. and Kolenbrander, G. V. (1965) Soil inorganic nitrogen. In *Soil Nitrogen, Agromony 10.* (eds W. V. Bartholomew and F. E. Clark), pp. 43–92. American Society of Agronomy, Madison, Wisconsin.

Harper, D. M. (1986) The effects of artificial enrichment upon the planktonic and benthic communities in a mesotrophic to hypertrophic loch series in lowland Scotland. *Hydrobiologia*, **137**, 9–19.

Harris, G. P. (1986) *Phytoplankton Ecology: Structure, Function, and Fluctuation.* Chapman and Hall, London. 384 pp.

Harrison, G. W. (1979) Stability under environmental stress: resistance, resilience, persistence, and variability. *Am. Nat.*, **113**, 659–69.

Harrison, G. W. and Fekete, S. (1980) Resistance of nutrient cycling systems to perturbations of the flow rates. *Ecol. Modell.*, **120**, 3–9.

Hartline, B. K. (1980) Research News. Coastal upwelling: Physical factors feed fish. *Science*, **208**, 38–40.

Harvey, H. W. (1963) The Chemistry and Fertility of Sea Waters. Cambridge University Press, Cambridge, England. 240 pp.

Harwell, M. A., Cropper, Jr., W. P. and Ragsdale, H. A. (1977) Nutrient recycling and stability: A reevaluation. *Ecology*, **58**, 660–6.

Harwell, M. A., Cropper, Jr., W. P. and Ragsdale, H.L. (1981) Analysis of transient characteristics of a nutrient cycling model. *Ecol. Modell.*, **12**, 105–31.

Hassall, M., Turner, J. G. and Rands, M. R. W. (1987) Effects of terrestrial isopods on the decomposition of woodland leaf litter. *Oecologia (Berlin)*, **72**, 597–604.

Hassell, M. P. (1978) *The Dynamics of Arthropod Predator–Prey Systems.* Monographs in Population Biology 13. Princeton University Press, Princeton, New Jersey.

Haukioja, E., Kapainen, K., Niemela, P. and Tuomi, J. (1983) Plant availability hypothesis and other explanations of herbivore cycles: Complementary or exclusive alternatives? *Oikos*, **40**, 419–32.

Hayes, F. R., McCarter, J. A., Cameron, M. L. and Livingstone, D. A. (1952) On the kinetics of phosphorus exchange in lakes. *J. Ecol.*, **39**, 202–16.

Headley, A. D., Callaghan, T. V. and Lee, J. A. (1985) The phosphorus economy of the evergreen tundra plant, *Lycopodium annotinum*. *Oikos*, **45**, 235–45.

Hecky, R. E. (1988) Nutrient limitation of phytoplankton in freshwater and marine environments: A review of recent evidence of the effects of enrichment. *Limnol. Oceanogr.*, **33**, 796–822.

Hecky, R. E. and Kilham, P. (1988) Nutrient limitation of phytoplankton in freshwater and marine environments: A review of recent evidence on the effects of enrichment. *Limnol. Oceanogr.*, **33**, 796–822.

Henderson, G. S. and Harris, W. F. (1975) An ecosystem approach to the characterization of the nitrogen cycle in a deciduous forest watershed. In *Forest Soils and Forest Land Management* (eds B. Bernier and C. H. Winget), Laval University Press, Quebec, P.Q., Canada.

Hendrix, S. D. (1984) Reactions of *Heracleum lanatum* to floral herbivory by *Depressaria pastinacella*. *Ecology*, **65**, 191–97.

Henry, R. L., III. (1985) The impact of zooplankton size structure on phosphorus cycling in field enclosures. *Hydrobiologia*, **120**, 3–9.

Hershey, A. E., Hiltner, A. L., Hullar, M. A. J., Miller, M. C., Vestal, J. R., Lock, M. A., Rundel, S. and Peterson, B. J. (1988) Nutrient influence on a stream grazer: *Orthocladius* microcommunities respond to nutrient input. *Ecology*, **69**, 1383–92.

Hessen, D. O. and Nilssen, J. P. (1986) From phytoplankton to detritus and bacteria: Effects of short-term nutrient and fish perturbations in a eutrophic lake. *Arch. Hydrobiol.*, **105**, 273–84.

Hilbert, D. W., Swift, D. M., Detling, J. K. and Dyer, M. I. (1979) Relative growth rates and the grazing optimization hypothesis. *Oecologia*, **51**, 14–18.

Holling, C. S. (1959) Some characteristics of simple types of predation and parasitism. *Can. Entomol.*, **91**, 385–98.

Holling, C. S. (1973) Resilience and stability in ecological systems. *Annu. Rev. Ecol. System.*, **4**, 1–23.

Hopkinson, C. S. and Shubauer, J. P. (1984) Static and dynamic aspects of nitrogen cycling in the salt marsh graminoid *Spartina alterniflora*. *Ecology*, **65**, 961–9.

Houghton, R. A., Schlesinger, W. H., Brown, S. and Richards, J. F. (1985). Carbon dioxide exchange between the atmosphere and terrestrial ecosystems. In *Atmospheric Carbon Dioxide and the Global Carbon Cycle* (ed. J. R. Trabalka), pp. 113–140. DOE/ER-0239. Carbon Dioxide Research Division, U.S. Department of Energy, Washington, D.C.

Howard-Williams, C. and Allanson, B. R. (1981). Phosphorus cycling in a dense *Potamogeton pectinatus* L. bed. *Oecologia (Berlin)*, **49**, 55–66.

Howarth, R. W. (1988) Nutrient limitation of net primary production in marine ecosystems. *Ann. Rev. Ecol. System.* **19**, 89–110.

Hrbacek, J., Dvorakova, M., Korinek, V. and Prochazkova, L. (1961) Demonstration of the effect of the fish stock on the species composition of zooplankton and the intensity of metabolism of the whole plankton assemblage. *Verh. Int., Verh. Theoret. Angew. Limnol.*, **14**, 192–5.

Hsu, S. B., Hubbell, S. P. and Waltman, P. (1977) A mathematical theory for single-nutrient competition in continuous cultures of microorganisms. *SIAM J. Appl. Mathemat.*, **32**, 366–83.

Huff, D. D., Koonce, J. F., Ivarson, W. R., Weiler, P. R., Dettmann, E. H. and Harris, R. F. (1973) Simulation of urban runoff, nutrient loading, and biotic response of a shallow eutrophic lake. In *Workshop Proceedings* (eds. E. J. Middlebrooks, D. H. Falkenborg and T. E. Malony) pp. 33–55. Utah State University, Logan, Utah.

Hurd, L. E. and Wolf, L. L. (1974) Stability in relation to nutrient enrichment in arthropod consumers of old-field successional ecosystems. *Ecol. Monogr.*, **44** 465–82.

Huston, M. A. (1979) A general hypothesis on species diversity. *Am. Nat.*, **113**, 81–101.

Hutchinson, G. E. (1948) Circular casual systems in ecology. *Ann. N.Y. Acad. Sci.*, **50**, 221–46.

Hutchinson, G. E. (1953) The concept of pattern in ecology. *Proc. Natl. Acad. Sci. Philadelphia*, **105**, 1–12.

Hutchinson, G. E. (1959) Homage to Santa Rosalia or why are there so many kinds of animals. *Am. Nat.*, **93**, 155–60.

Hutchinson, G. E. (1961) The paradox of the plankton. *Am. Nat.*, **95**, 137–45.

Hutchinson, G. E. (1964) The lacustrine microcosm reconsidered. *Am. Sci.*, **52**, 334–41.

Imboden, D. M. (1974) Phosphorus model of lake eutrophication. *Limnol. Oceanogr.*, **19**, 297–304.

Ingham, E. R., Trofymow, J. A., Ames, R. N., Hunt, H. W., Morley, C. R., Moore, J. C. and Coleman, D. C. (1986) Trophic interactions and nitrogen cycling in a semi-arid grassland soil. I. Seasonal dynamics of the natural populations, their interactions and effects on nitrogen cycling. *J. Appl. Ecol.*, **23**, 597–614.

Ingham, R. E., Trofymow, J. A., Ingham, E. R. and Cileman, D. C. (1985) Interactions of bacteria, fungi, and their nematode grazers: Effects of nutrient cycling and plant growth. *Ecol. Monogr.*, **55**, 119–40.

Inouye, D. W. (1978) Resource partitioning in bumblebees: Experimental studies of foraging behavior. *Ecology*, **59**, 672–8.

Inouye, R. S. (1987) Pocket gophers (*Geomys bursarius*), vegetation, and soil nitrogen along a successional sere in east central Minnesota. *Oecologia (Berlin)*, **72**, 178–84.

Inouye, R. S., Huntley, N. J., Tilman, D., Tester, J. R. (1987) Pocket gophers (*Geomys bursorius*), vegetation and soil nitrogen along a successional sere in east central Minnesota. *Oecologia*, **72**, 178–84.

Jacquez, J. A. (1972) *Compartmental Analysis in Biology and Medicine*. Elsevier Publishing Company, Amsterdam. 237 pp.

Janos, D. P. (1983). Tropical mycorrhizas, nutrient cycles, and plant growth. In

Tropical Rain Forest: Ecology and Management (eds S. L. Sutton, T. C. Whitmore and A. C. Chadwick), pp. 327–345. Blackwell Scientific, Oxford.

Janus, L. L. and Vollenweider, R. A. (1981) *The OECD Cooperation programme on eutroplucation* (*Summary report*, Canadian Contribution) Scientific Series No. 131. National Water Research Institute, Inland Waters Directorate, Canada Centre for Inland Waters, Burlington, Ontario.

Jenny, H. (1980) *The Soil Resource: Origin and Behavior. Ecological Studies*, Vol. 37. Springer-Verlag, New York. 377 pp.

Johannes, R. E. (1965) Influence of marine protozoa on marine regeneration. *Limnol. Oceanogr.*, **10**, 434–42.

Johannes, R. E. (1968) Nutrient regeneration in lakes and oceans. *Adv. Microbiol. Sea*, **1**, 203–13.

Johnson, C. W. (1985) *Bogs of the Northeast*. University Press of New England, Hanover, New Hampshire. 269 pp.

Jordan, C. F. (1985) *Nutrient Cycling in Tropical Rain Forests*. John Wiley and Sons, New York. 190 pp.

Jordan, C. F. and Herrera, R. (1981) Tropical rain forests: are nutrients really critical? *Am. Nat.*, **117**, 167–80.

Jordan, C. F., Kline, J. R. and Sasscer, D. S. (1972) Relative stability of mineral cycles in forest ecosystems. *Am. Nat.*, **106**, 237–53.

Jorgensen, S. E., Kamp-Nielsen, L. and Jacobsen, O. S. (1975) A submodel of anaerobic mud-water exchange of phosphate. *Ecol. Modell.*, **1**, 133–46.

Jost, J. L., Drake, J. F., Tsuchiya, H. M. and Fredrickson, A. G. (1973) Microbial food chains and food webs. *J. Theor. Biol.*, **41**, 461–84.

Juday, C., Rich, W. H., Kemmerer, G. I. and Mann, A. (1932) Limnological studies of Karluk Lake, Alaska, 1926–1930. *U.S. Bur. Fish. Bull.* **57**, 407–36.

Kachi, N. and Hirose, T. (1983) Limiting nutrients for plant growth in coastal sand dune soils. *J. Ecol.*, **71**, 937–44.

Kalela, O. (1962) On the fluctuations in the numbers of arctic and boreal small rodents as a problem of production ecology. *Ann. Acad. Sci. Fenn. Ser. A, IV*, **66**, 1–38.

Kalff, J. (1983) Phosphorus limitation in some tropical African lakes. *Hydrobiologia*, **100**, 101–12.

Kelly, J. R. and Levin, S. A. (1986) A comparison of aquatic and terrestrial nutrient cycling and production processes in natural ecosystems, with reference to ecological concepts of relevance to some waste disposal issues. In *The Role of the Oceans as a Waste Disposal Option* (ed. G. Kullenberg), pp. 165–203. D. Reidel Publishing Company, New-York.

Kelly, J. M. and Beauchamp, J. J. (1987) Mass loss and nutrient changes in decomposing upland oak and mesic mixed-hardwood leaf litter. *Soil Sci. Soc. Am. J.*, **51**, 1616–22.

Kemp, W. M. and Mitsch, W. J. (1979) Turbulence and phytoplankton diversity: A general model of the 'paradox of plankton.' *Ecol. Modell.* **7**, 201–22.

Kerfoot, W. C. and DeMott, W. R. (1984) Food web dynamics: Dependent chains and vaulting. In *Trophic Interactions Within Aquatic Ecosystems* (eds D. G. Meyers and J. R. Strickler), pp. 347–382. American Association for the Advancement of Science Selected Symposium.

Kerfoot, W. C., Levitan, C. and DeMott, W. R. (1988) *Daphnia*–phytoplankton interactions: Density-dependent shifts in resource quality. *Ecology*, **69**, 1806–25.

Ketchum, B. H. (1939) The development and restoration of deficiencies in the phosphorus and nitrogen composition of unicellular plants. *J. Cell Comp. Physiol.*, **13**, 373–81.

Khanna, P. K. and Ulrich, B. (1981) Changes in chemistry of throughfall under stands of beech and spruce following the addition of fertilizers. *Acta Oecol., Oecol. Plant.*, **2**, 155–64.

Kiefer, D. A. and Atkinson, C. A. (1984) Cycling of nitrogen by plankton: A hypothetical description based upon efficiency of energy conversion. *J. Mar. Res.*, **42**, 655–75.

Kimmins, J. P. (1972) Relative contributions of leaching, litter fall and defoliation by *Neodiprion sertifer* (Hymenoptera) to the removal of cesium-134 from red pine. *Oikos*, **23**, 226–34.

Kirchner, W. B. and Dillon, P. J. (1975) An empirical method of estimating the retention of phosphorus in lakes. *Water Resour. Res.*, **11**, 182–3.

Kitchell, J. F., O'Neill, R. V., Webb, D., Gallepp, G. W., Bartell, S. M., Koonce, J. F. and Ausmus, B. S. (1979) Consumer regulation of nutrient cycling. *BioScience*, **29**, 28–34.

Kliejunas, J. T. and Ko, W. H. (1974) Deficiency of inorganic nutrients as a contributing factor to ohi'a decline. *Phytopathology*, **64**, 891–6.

Koide, R. T., Huennecke, L. F. and Mooney, H. A. (1987) Gopher mound soil reduces growth and affects ion uptake of two annual grassland species. *Oecologia (Berlin)*, **72**, 284–90.

Kremer, J. N. and Nixon, S. W. (1978) *A Coastal Marine Ecosystem*. Springer-Verlag, Berlin. 217 pp.

Krokhin, E. M. (1967) Effect of size of escapement of sockeye salmon spawners on the phosphate content of a nursery lake. Izvestiya Tikhookeanskogo Nauchno-Issledovatel'skogo. Instituta Rybnogo Kjozyaistva i Okeanografii. 57. Fisheries Research Board of Canada. Translation Series No. 1186, 31–54.

Laine, K. and Henttonen, H. (1982) The role of plant production in microtine cycles in northern Fennoscandia. *Oikos*, **40**, 407–18.

Lang, G. E. and Forman, R. T. T. (1978) Detrital dynamics in a mature oak forest: Hutcheson Memorial Forest, New Jersey. *Ecology*, **59**, 580–95.

Lasker, R. (1989) In *Toward a Theory of Biological–Physical Interaction in the World Ocean* (ed. B. J. Rothschild), Kluwer, Dordrecht.

Lauenroth, W. K. (1979) Grassland primary production: North American grasslands in perspective. In *Perspectives in Grassland Ecology* (ed. N. R. French), pp. 3–24. Springer-Verlag, New York.

Lehman, J. T. (1976) The filter feeder as an optimal forager and the predicted shapes of feeding curves. *Limnol. Oceanogr.*, **21**, 501–16.

Lehman, J. T. (1978) Aspects of nutrient dynamics in freshwater communities. Ph. D. thesis, University of Washington. 180 pp.

Lehman, J. T. (1980) Release and cycling of nutrients between planktonic algae and herbivores. *Limnol. Oceanogr.*, **25**, 620–32.

Lehman, J. T., Botkin, D.B. and Likens, G. E. (1975) The assumptions and rationales of a computer model of phytoplankton population dynamics. *Limnol. Oceanogr.*, **20**, 343–64.

Levin, S. A. (1974) Dispersion and population interactions. *Am. Nat.*, **108**, 207–28.

Levin, S. A. (1976) Population dynamic models in heterogeneous environments. *Annu. Rev. Ecol. System.*, **7**, 287–310.

Levin, S. A. and Paine, R. T. (1974) Disturbance, patch formation, and community structure. *Proc. Natl. Acad. Sci. U.S.A.*, **71**, 2744–7.

Levine, S. H. (1976) Competitive interactions in ecosystems. *Am. Nat.*, **110**, 903–10.

Levins, R. (1979) Coexistence in a variable environment. *Am. Nat.*, **114**, 765–83.

Levitan, C., Kerfoot, W. C. and DeMott, W. R. (1985) Ability of *Daphnia* to buffer trout lakes against periodic nutrient inputs. *Verh. Int. Verein. Limnol.*, **22**, 3076–82.

Lewis, M. R., Harrison, W. G., Oakey, N. S., Hebert, D. and Platt, T. (1986) Vertical nitrate fluxes in the oligotrophic ocean. *Science*, **234**, 870–3.

Likens, G. E. (ed.) (1985) *An Ecosystem Approach to Aquatic Ecology: Mirror Lake and Its Environment.* Springer-Verlag, New York. 516 pp.

Lindeman, R. L. (1942) The trophic-dynamic aspect of ecology. *Ecology*, **23**, 399–418.

Lorenzen, M. W., Smith, D. J. and Kimmel, L. V. (1976) A long-term phosphorus model for lakes: Application to Lake Washington. In *Modeling Biochemical Processes in Aquatic Ecosystems* (ed. R. P. Canale), pp. 75–92. Ann Arbor Science Publishers, Ann Arbor, Michigan.

Lotka, A. J. (1956) *Elements of Mathematical Biology*. Dover Publications, New York, 465 pp. (Originally published in 1924 by Williams and Wilkins Co.)

Lovstad, O. (1984) Growth limiting factors for *Oscillatoria agardhii* and diatoms in eutrophic lakes. *Oikos*, **42**, 185–92.

Luckinbill, L. S. (1974) Coexistence in laboratory populations of *Paramecium aurelia* and its predator *Didinium nasutum*. *Ecology*, **55**, 1142–7.

Luckinbill, L. S. (1979) Rgulation, stability, and diversity in a model experimental microcosm. *Ecology*, **60**, 1098–102.

Ludwig, D., Jones, D. D. and Holling, C. S. (1978) Qualitative analysis of insect outbreak systems: the spruce budworm and forest. *J. Anim. Ecol.*, **47**, 315–32.

Luxmoore, R. J. (1989) Modeling chemical transport, uptake, and effects in the soil–plant–litter system. In *Analysis of Biogeochemical Cycling Processes in Walker Branch Watershed* (eds D. W. Johnson and R. I. Van Hook) pp. 352–384. Springer-Verlag, New York.

Luxmoore, R. J., Begovich, C. L. and Dixon, K. R. (1978) Modelling solute uptake and incorporation into vegetation and litter. *Ecol. Modell.*, **5**, 137–71.

Lynch, M. and Shapiro, J. (1981) Predation, enrichment, and phytoplankton community structure. *Limnol. Oceanogr.*, **26**, 86–102.

MacArthur, R. H. and Connell, J. H. (1966) *The Biology of Populations*. John Wiley, New York.

Madeira, P. T., Brooks, A. S. and Seale, D. B. (1982) Excretion of total phosphorus, dissolved reactive phosphorus, ammonia, and urea by Lake Michigan *Mysis relicta*. *Hydrobiologia*, **93**, 145–54.

Maguire, B., Jr., Slobodkin, L. B., Morowitz, H. J., Moore, III, B. and Botkin, D. B. (1980) A new paradigm for the examination of closed ecosystems. In *Microcosms in Ecological Research*. (ed. J. P. Giesey), pp. 30–68. CONF-781101, Technical Information Center. Springfield, Virginia. USDOE.

Malone, T. C. (1980) Algal size. In *Physiological Ecology of Phytoplankton* (ed. I. Morris) pp. 433–464. University of California Press, Berkeley, California.

Marinucci, A. C., Hobbie, J. E. and Helfrich, J. V. K. (1983) Effect of litter nitrogen on decomposition and microbial biomass of *Spartina alterniflora*. *Microb. Ecol.*, **9**, 27–40.

Martin, J. H. (1967) Phytoplankton–zooplankton relationships in Narragansett Bay. III. Seasonal changes in zooplankton excretion rates in relation to phytoplankton abundance. *Limnol. Oceanogr.*, **13**, 63–71.

Martin, J. H., Gordon, R. M. and Fitzwater, S. E. (1990) Iron in Antarctic waters. *Nature*, **345**, 156–8.

Matson, P. A. and Vitousek, P. M. (1981) Nitrification potentials following clearcutting in the Hoosier National Forest, Indiana. *For. Sci.*, **27**, 781–91.

Mattson, W. J. and Addy, N. D. (1975) Phytophagous insects as regulators of forest primary production. *Science*, **190**, 515–22.

Mattson, W. J. (1980) Herbivory in relation to plant nitrogen content. *Ann. Rev. Ecol. System.* **11**, 119–61.

May, R. M. (1972) Limit cycles in predator–prey communities. *Science*, **177**, 900–2.

May, R. M. (1973) *Stability and Connectance in Model Ecosystems. Monographs in Population Biology*, vol. 6. Princeton University Press, Princeton, New Jersey. 235 pp.

May, R. M. (ed.) (1976) *Theoretical Ecology: Principles and Applications.* W. B. Saunders and Company, Philadelphia. 317 pp.

Maynard Smith, J. (1975) *Models in Ecology.* Cambridge University Press, Cambridge. 146 pp.

McCarthy, J. J. and Altabet, M. A. (1984) Patchiness in nutrient supply: implications for phytoplankton ecology, In *Trophic Interactions within Aquatic Ecosystems* (eds D. G. Meyers and J. R. Strickler), pp. 29–47. American Association for the Advancement of Science Selected Symposium. 85. Westview Press, Boulder, Colorado.

McCarthy, J. J. and Goldman, J. C. (1979) Nitrogen nutrition of marine phytoplankton in nutrient depleted waters. *Science*, **203**, 670–72.

McIntosh, R. I. (1985) *The Background of Ecology.* Cambridge University Press, Cambridge, 383 pp.

McMurtrie, R. E. (1986) Forest productivity in relation to carbon partitioning and nutrient cycling: A mathematical model. In *Attributes of Trees as Crop Plants*, (eds Connell M. G. R. and Jackson, J. E.), pp. 194–207. Institute of Terrestrial Ecology, Huntingdon, England.

McNaught, D. C. and Scavia, D. (1976) Application of a model of zooplankton composition to problems of fish introduction to the Great Lakes. In *Modeling of Biochemical Processes in Aquatic Ecosystems* (ed. R. P. Canale), pp. 281–304. Ann Arbor Science Publishers, Ann Arbor, Michigan.

McNaughton, S. J. (1979a) Grazing as an optimization process: Grass ungulate relationships in the Serengeti. *Am. Nat.*, **113**, 691–703.

McNaughton, S. J. (1979b) Grassland–herbivore dynamics. In *Serengeti: Studies of Ecosystem Dynamics in a Tropical Savanna* (eds A. R. E. Sinclair and M. Norton-Griffiths), pp. 46–81. University of Chicago Press, Chicago, Illinois.

McNaughton, S. J. (1984) Grazing lawns: Animals in herds, plant form, and coevolution. *Am. Nat.*, **124**, 863–86.

McNaughton, S. J. and Chapin, III, F. S. (1985) Effects of phosphorus nutrition and defoliation on C_4 graminoids from the Serengeti plains. *Ecology*, **66**, 1617–29.

McNaughton, S. J., Wallace, L. and Coughenour, M. B. (1983) Plant adaptation in an ecosystem context: Effect of defoliation, nitrogen, and water on growth of an African C_4 sedge. *Ecology*, **64**, 307–18.

McQueen, D. J., Post, J. R. and Mills, E. L. (1986) Trophic relationships in freshwater pelagic ecosystems. *Can. J. Fish. Aquat. Sci.*, **43**, 1571–81.

Mech, L. D., Frenzel, Sr., L. D., Ream, R. R. and Winship, J. W. (1971) Movements, behavior, and ecology of timber wolves in northeastern Minnesota. In *Ecological Studies of the Timber Wolf in Northeastern Minnesota* (eds L. D. Mech and L. D. Frenzel, Jr.), USDA Forest Service Research Paper NC-52.

Mellinger, M. V. and McNaughton, S. J. (1975) Structure and function of successional vascular plant communities in central New York. *Ecol. Monogr.*, **45**, 161–82.

Mills, E. L. and Schiavone, Jr., A. (1982) Evaluation of fish communities through assessment of zooplankton populations and measures of lake productivity. *N. Am. J. Fish. Manag.*, **2**, 14–27.

Mills, E. L., Forney, J. L. and Wagner, K. J. (1987) Fish predation and its cascading effect in the Oneida Lake food chain. In *Predation: Direct and Indirect Impacts on Aquatic Communities* (eds W. C. Kerfoot and A. Sih), pp. 118–131. University Press of New England, Hanover, New Hampshire.

Moran, N. and Hamilton, W. D. (1980) Low nutritive quality as defense against herbivores. *J. Theor. Biol.*, **86**, 247–54.

Morowitz, H. J. (1968) *Energy Flow in Biology*. Academic Press, New York.

Morris, J. T. and Lajtha, K. (1986) Decomposition and nutrient dynamics of litter from four species of freshwater emergent macrophytes. *Hydrobiologia*, **131**, 215–23.

Morris, D. P. and Lewis, Jr., W. M. (1988) Phytoplankton nutrient limitation in Colorado lakes. *Freshw. Biol.*, **20**, 315–27.

Mortimer, C. H. (1942) The exchange of dissolved substances between mud and water in lakes. *J. Ecol.*, **30**, 147–201.

Mulholland, P. J., Newbold, J. D., Elwood, J. W. and Hom, C. L. (1983) The effect of grazing intensity on phosphorus spiralling in autotrophic streams. *Oecologia (Berlin)*, **58**, 358–66.

Murata, Y. (1977) *Mathematics for Stability and Optimization of Economic Systems*. Academic Press, New York. 418 pp.

Murdoch, W. W. and Oaten, A. (1975) Predation and population stability. *Adv. Ecol. Res.*, **9**, 1–131.

Myers, J. H. (1979) The effects of food quantity and quality on emergence time in the cinnabar moth. *Can. J. Zool.*, **52**, 1150–6.

Myers, J. H. (1980) Is the insect or the plant the driving force in the cinnabar moth tansy ragwort system? *Oecologia (Berlin)*, **47**, 16–21.

Myers, J. H. and Post, B. J. (1981) Plant nitrogen and fluctuations of insect populations: A test with the cinnabar moth–tansy ragwort system. *Oecologia (Berlin)*, **48**, 151–6.

Neftel, A., Moor, E., Oeschger, H. and Stauffer, B. (1985) Evidence from polar ice cores for the increase in atmospheric CO_2 in the past two centuries. *Nature*, **315**, 45–7.

Neill, W. E. (1984) Regulation of rotifer densities by crustacean zooplankton in an oligotrophic montane lake in British Columbia. *Oecologia (Berlin)*, **61**, 175–81.

Neill, W. E. (1988) Complex interactions in oligotrophic lake food webs: Responses to nutrient enrichment. In *Complex Interactions in Lake Communities* (ed. S. R. Carpenter), pp. 31–44. Springer-Verlag, New York.

Newbold, J. D., Elwood, J. W., O'Neill, R. V. and Van Winkle, W. (1981) Measuring nutrient spiralling in streams. *Can. J. Fish. Aquat. Sci.*, **38**, 860–3.

Newbold, J. D., O'Neill, R. V., Elwood, J. W. and Van Winkle, W. (1982) Nutrient spiralling in streams: Implications for nutrient limitation and invertebrate activity. *Am. Nat.*, **120**, 628–52.

Nisbet, R. M. and Gurney, W. S. C. (1976) Model of material cycling in a closed ecosystem. *Nature*, **264**, 633–4.

Nisbet, R. M., McKinistry, J. and Gurney, W. S. C. (1983) A 'strategic' model of material cycling in closed ecosystem. *Mathemat. Biosci.*, **64**, 99–113.

Noy-Meir, I. (1975) Stability of grazing systems: An application of predator–prey graphs. *J. Ecol.*, **63**, 459–81.

Nyholm, N. (1978) A simulation model for phytoplankton growth and nutrient cycling in eutrophic, shallow lakes. *Ecol. Modell.*, **4**, 279–310.

Nunney, L. (1980) The stability of complex model ecosystems. *Am. Nat.*, **111**, 515–25.

Nürnberg, G. K. (1984) The prediction of internal phosphorus load in lakes with anoxic hypolimnia. *Limnol. Oceanogr.*, **29**, 111–24.

O'Brien, W. J. (1974) The dynamics of nutrient limitation of phytoplankton: A model reconsidered. *Ecology*, **55**, 135–41.

O'Connor, D. J. and Mueller, J. A. (1970) A water quality model of chlorides in Great Lakes. *Proc. Am. Soc. Civ. Eng. J. San. Eng. Div.*, **96**, 955–75.

Odum, E. P. (1969) The strategy of ecosystem development. *Science*, **164**, 262–70.

Odum, E. P. (1971) *Fundamentals of Ecology*, 3rd edn. W. B. Saunders Company, Philadelphia, Pennsylvania.

Odum, H. T. (1957) Primary production measurements in eleven Florida springs and a marine turtle grass community. *Limnol. Oceanogr.*, **2**, 85–97.

Odum, H. T. and Odum, E. C. (1976) *Energy Basis for Man and Nature*. McGraw-Hill Book Company, New York. 337 pp.

Odum, H. T. and Odum, E. P. (1955) Trophic structure and productivity of a windward coral reef community on Eniwetok Atoll. *Ecol. Monogr.*, **25**, 291–320.

Odum, H. T. and Pigeon, R. F. (eds) (1970) *A Tropical Rain Forest*. USAEC, Division of Technical Information, Washington, D.C.

Odum, W. E. (1970) Pathways of energy flow in a south Florida estuary. Ph. D. dissertation. University of Miami. 180 pp.

Oechel, W. and Strain, B. R. (1985) Native species responses to increased carbon dioxide concentration. In *Direct Effects of Increasing Carbon Dioxide in Vegetation* (eds B. R. Strain and J. D. Cure). DOE/ER-0238.

Officer, C. B., Biggs, R. B., Taft, J. L., Cronin, L. E., Taylor, M. A. and Boynton, W. R. (1984) Chesapeake Bay anoxia: Origin, development, and significance. *Science*, **223**, 22–7.

Oksanen, L. S., Fretwell, D., Arruda, J. A. and Niemela, P. (1981) Exploitation ecosystems in gradients of primary productivity. *Am. Nat.*, **118**, 240–61.

Okubo, A. (1968) Some remarks on the importance of the 'shear effect' on horizontal diffusion. *J. Oceanogr. Soc. Japan*, **24**, 60–9.

Okubo, A. (1974) *Diffusion-induced instability in model ecosystems*. Chesapeake Bay Institute. The John Hopkins University, Baltimore. Tech. Rept., **86**.

Okubo, A. (1978) Horizontal dispersion and critical scales of phytoplankton patches in Spatial Pattern in Plankton Communities (ed. Steele, J. H.) Plenum Press, New York, 21–42 pp.

Okubo, A. (1980) *Diffusion and Ecological Problems: Mathematical Models*. Springer-Verlag, New York.

Oliver, J. D. and Legovic, T. (1988) Okefenokee marshland before, during and after nutrient enrichment by a bird rookery. *Ecol. Modell.*, **43**, 195–224.

Olsen, P. and Willen, E. (1980) Phytoplankton response to a sewage reduction in Vattern, a large, oligotrophic lake in Central Sweden. *Arch. Hydrobiol.*, **89**, 171–88.

Olson, J. S. (1965) Equations for cesium transfer in a *Liriodendron* forest. *Hlth. Phys.*, **11**, 1385–92.

Olson, J. S., Watts, J. A. and Allison, L. J. (1983) Carbon in live vegetation of major world ecosystems. ORNL-5862, Oak Ridge National Laboratory, Oak Ridge, Tennessee.

O'Neill, R. V. (1976) Ecosystem persistence and heterotrophic regulation. *Ecology*, **57**, 1244–53.

O'Neill, R. V., DeAngelis, D. L., Pastor, J. J., Jackson, B. J. and Post, W. M. (1989) Multiple nutrient limitations in ecological models. *Ecol. Modell.*, **46** 147–63.

O'Neill, R. V., DeAngelis, D. L., Waide, J. B. and Allen, T. F. H. (1986) *A Hierarchical Concept of the Ecosystem*. Princeton University Press, Princeton, New Jersey. 253 pp.

O'Neill, R. V., Elwood, J. W. and Hildebrand, S. G. (1979) Theoretical implications of spatial heterogeneity in stream ecosystems. In *Systems Analysis of Ecosystems* (eds G. S. Innis and R. V. O'Neill), pp. 79–101. International Cooperative Publishing House, Fairland, Maryland.

O'Neill, R. V. and Reichle, D. E. (1980) Dimensions of ecosystem theory. In *Forests: Fresh Perspectives from Ecosystem Analysis. Fortieth Annual Biology Colloquium*, pp. 11–26. Oregon State University Press, Corvallis, Oregon.

Owen, D. F. (1980) How plants may benefit from the animals that eat them. *Oikos*, **35**, 230–5.

Pace, M. L. (1984) Zooplankton community structure, but not biomass, influences the phosphorus–chlorophyll *a* relationship. *Can. J. Fish. Aquat. Sci.*, **41**, 1089–96.

Paige, K. N. and Whitham, T. G. (1987) Overcompensation in response to mammalian herbivory: the advantage of being eaten. *Am. Nat.*, **129**, 406–16.

Paine, R. T. (1966) Food web complexity and species diversity. *Am. Nat.*, **100**, 65–75.

Park, T. (1954) Experimental studies of interspecific competition. II. Temperature, humidity and competition in two species of *Tribolium*. *Physiol. Zool.*, **27**, 177–238.

Parker, R. A. (1978) Nutrient recycling in closed ecosystem models. *Ecol. Modell.*, **4**, 67–70.

Parnas, H. (1975) Model for decomposition of organic material by microorganisms. *Biol. Biochem.*, **7**, 161–9.

Parsons, T., McAllister, C. D., LaBrasseur, R. J. and Barraclough, W. E. (1970) The use of nutrients in the enrichment of sockeye salmon nursery lakes. (Paper given at the thirty-third meeting of the American Society of Limnology and Oceanography, Kingston, Rhode Island.)

Pastor, J., Aber, J. D., McClaugherty, C. A. and Melillo, J. M. (1984) Aboveground production and N and P cycling along a nitrogen mineralization gradient on Blackhawk Island, Wisconsin. *Ecology*, **65**, 256–68.

Pastor, J. J. and Post, W. M. (1986) Influence of climate, soil moisture and succession on forest carbon and nitrogen cycles. *Biogeochemistry*, **2**, 3–27.

Pastor, J. J. and Post, W. M. (1988) Response of northern forests to CO_2-induced climate change. *Nature*, **334**, 55–8.

Patalas, K. (1972) Crustacean plankton and the eutrophication of St. Lawrence Great Lakes. *J. Fish. Res. Bd. Can.*, **29**, 1451–62.

Patten, B. C. (1971) A primer for ecological modeling and simulation with analog and digital computers. In *Systems Analysis and Simulation in Ecology* (ed. B. C. Patten), Vol. 4, pp. 3–121. Academic Press, New York.

Peng, T.-H. (1986) Land use change and carbon exchange in the tropics: II. Estimates for the entire region: *comment*. *Environ. Manag.*, **10**, 573–5.

Penning de Vries, F. W. T., Murphy, Jr., C. E., Wells, C. G. and Jorgensen, J. R. (1974) Simulation of nitrogen distribution in time and space in even-aged loblolly pine plantations and its effect on productivity. IBP Contribution No. 156. *Proc. Symposium on Mineral Cycling in Southeastern Ecosystems*, Augusta, Georgia, 1–3 May, 1974.

Perry, D. A., Amaranthus, M. P., Borchers, J. G., Borchers, S. L. and Brainerd, R. E. (1989) Bootstrapping in ecosystems. *BioScience*, **39**, 230–7.

Peterjohn, W. T. and Correll, D. L. (1984) Nutrient dynamics in an agricultural watershed: Observations on the role of a riparian forest. *Ecology*, **65**, 1466–75.

Peters, R. H. (1975) Phosphorus excretion and the measurement of feeding and assimilation by zooplankton. *Limnol. Oceanogr.*, **20**, 858–9.

Peters, R. H. and Rigler, F H. (1973) Phosphorus release by *Daphnia*. *Limnol. Oceanogr.*, **18**, 821–39.

Peterson, D. L. and Rolfe, G. L. (1982) Nutrient dynamics and decomposition of litterfall in flood plain and upland forests of central Illinois. *Forest Sci.*, **28**, 667–81.

Peterson, B. J., Hibbie, J. E., Hershey, A. E., Lock, M. A., Fird, T. E., Vestal, J. R., Hullar, M. A. J., Miller, M. C., Ventullo, R. M. and Volk, G. S. (1985) Transformation of a tundra river from heterotrophy to autotrophy by addition of phosphorus. *Science*, **229**, 138–6.

Peterson, R. C. and Cummins, K. W. (1974) Leaf processing in a woodland stream. *Freshw. Biol.*, **4**, 343–68.

Phillipson, J. (1966) *Ecological Energetics. Studies in Biology* No. 1., Edward Arnold Publishers, London. 57 pp.

Pickett, S. T. A. and White, P. S. (eds) (1985) *The Ecology of Natural Disturbance and Patch Dynamics.* Academic Press, Orlando, Florida. 472 pp.

Pimm, S. L. (1979) The structure of food webs. *Theor. Pop. Biol.*, **16**, 148–58.

Pimm, S. L. (1980a) Bounds on food web connectance. *Nature*, **284**, 591.

Pimm, S. L. (1980b) Properties of food webs. *Ecology*, **61**, 219–25.

Pimm, S. L. (1982) *Food Webs.* Chapman and Hall, Publishers, London. 219 pp.

Pimm, S. L. and Lawton, J. H. (1977) The number of trophic levels in ecological communities. *Nature*, **268**, 329–31.

Pimm, S. L. and Lawton, J. H. (1978) On feeding on more than one trophic level. *Nature*, **275**, 542–4.

Pitelka, F. A. (1964) The nutrient-recovery hypothesis for arctic microtine rodents, In *Grazing in Terrestrial and Marine Environments* (ed. P. Crisp), pp. 55–68. Blackwell, Oxford.

Pomeroy, L. R. (1970) The strategy of mineral cycling. *Annu. Rev. Ecol. System.*, **1**, 171–90.

Porter, K. G. (1976) Enhancement of algal growth and productivity by grazing zooplankton. *Science*, **192**, 1332–3.

Porter, K. G. (1977) The animal–plant interface in freshwater ecosystems. *Am. Sci.*, **65**, 159–70.

Post, W. M., Peng, T.-S., Emanuel, R., King, W. A., Dale, V. H. and DeAngelis, D. L. (1990) The global carbon cycle. *An. Sci.*, **78**, 310–26.

Powell, T. and Richerson, P. J. (1985) Temporal variation, spatial heterogeneity, and competition for resources in plankton systems: a theoretical model. *Am. Nat.*, **125**, 431–63.

Prins, H. H. H., Ydenberg, Th., R. C. and Drent, R. H. (1980) The interaction of brent geese (*Branta bernicla*) and sea plaintain (*Plantago maritima*) during spring staging: Field observations and experiments. *Acta Bot. Neerland.*, **29**, 585–96.

Rafes, P. M. (1970) Insect influences on the forest canopy. In *Ecological Studies 1* (ed. D. E. Reichle). Springer-Verlag, Berlin.

Reichle, D. E., O'Neill, R. V. and Olson, J. S. (eds) (1973) *Modeling Forest Ecosystems.* EDFB/IBP-73/7. Oak Ridge National Laboratory, Oak Ridge, Tennessee.

Reiners, W. A. (1986). Complementary models for ecosystems. *Am. Nat.*, **127**, 59–73.

Reinertsen, H., Jensen, A., Langeland, A. and Olsen, Y. (1986) Algal competition for phosphorus: The influence of zooplankton and fish. *Can. J. Fish. Aquat. Sci.*, **43**, 1135–41.

Reuss, J. O. and Innis, G. S. (1977) A grassland nitrogen flow simulation model. *Ecology*, **58**, 379–88.

Rhee, G.-Y. (1973) A continuous culture study of phosphate uptake, growth rate and polyphosphate in *Scenedesmus* sp. *J. Phycol.*, **9**, 495–506.

Richerson, P. J., Armstrong, R. and Goldman, C. R. (1970) Contemporaneous disequilibrium: a new hypothesis to explain the 'paradox of plankton'. *Proc. Natl. Acad. Sci. U.S.A.*, **67**, 1710–14.

Richman, S., Bohon, S. A. and Robbins, S. E. (1980) Grazing interactions among freshwater calanoid copepods. pp. 219–233. In *Evolution and Ecology of Zooplankton Communities* (ed. W. C. Kerfoot) pp. 219–233. University Press of New England, Hanover, New Hampshire.

Riebesell, J. (1974). Paradox of enrichment: Destabilization of exploitation ecosystems in ecological time. *Science*, **171**, 385–87.

Rigler, R. H. (1973) A dynamic view of the phosphorus cycle in lakes. In *The Environmental Phosphorus Handbook* (eds E. J. Griffith *et al.*), John Wiley & Sons, New York, pp. 539–568.

Rigler, F. H. (1978) Passage of phosphorus through a catchment. In *Environmental Biogeochemistry and Geomicrobiology. Vol. 1, The Aquatic Environment* (ed. W. E. Krumbein), pp. 65–81. Ann Arbor Science Publishers, Ann Arbor, Michigan.

Riley, G. A., Stommel, H. and Bumpus, D. F. (1949) Quantitative ecology of the plankton of the Western North Atlantic. *Bull. Binghan Oceanogr. Coll.*, **12**, 1–169.

Riley, G. A. (1963) Theory of food-chain relations in the ocean, In *The Sea Vol. 2* (ed. M. N. Hill), Wiley Inter-Science, New York pp. 438–463.

Robarts, R. D. and Southall, G. C. (1977) Nutrient limitation of phytoplankton growth in seven tropical man-made lakes, with special reference to Lake McIlwaine, Rhodesia. *Arch. Hydrobiol.*, **79**, 1–35.

Rosenzweig, M. L. (1971) Paradox of enrichment: Destabilization of exploitation ecosystems in ecological time. *Science*, **171**, 385–7.

Rosenzweig, M. L. and MacArthur, R. H. (1963) Graphical representation and stability conditions of prey–predator interactions. *Am. Nat.*, **97**, 209–23.

Ruess, R. W. (1984) Nutrient movement and grazing: experimental effects of clipping and nitrogen source on nutrient uptake in *Kyllinga nervosa*. *Oikos*, **43**, 183–8.

Ruess, R. W. and McNaughton, S. J. (1987) Grazing and the dynamics of nutrient and energy regulated microbial processes in the Serengeti grasslands. *Oikos*, **49**, 101–10.

Ruess, R. W., McNaughton, S. J. and Coughenour, M. B. (1983) The effects of clipping, nitrogen source and nitrogen concentration on the growth response and nitrogen uptake of an East African sedge. *Oecologia (Berlin)*, **59**, 253–61.

Sakshaug, E. and Olsen, Y. (1986) Nutrient status of phytoplankton blooms in Norwegian waters and algal strategies for nutrient competition. *Can. J. Fish. Aquat. Sci.*, **43**, 389–96.

Sarmiento, J. L., Herbert, T. and Toggweiler, J. R. (1988) Mediterranean nutrient balance and episodes of anoxia. *Glob. Geochem. Cycles*, **2**, 427–44.

Satchell, J. E. (1974) Introduction. Litter-interface of animate/inanimate matter. In *Biology of Plant Litter Decomposition*, (eds C. H. Dickenson and G. J. F. Pugh) Vol. 1, pp. 13–44. Academic Press, London.

Schaefer, D. A. and Whitford, W. G. (1981) Nutrient cycling by the subterranean termite *Gnathamitermes tubiformes* in a Chihuahan desert ecosystem. *Oecologia (Berlin)*, **48**, 277–83.

Schindler, D. W. (1977) Evolution of phosphorus limitation in lakes. *Science*, **195**, 260–2.

Schindler, D. W., Armstrong, F. A. J., Holmgren, S. K. and Brunskill, G. F. (1971) Eutrophication of lake 227, Experimental lakes areas, Northwestern Ontario by addition of phosphate and nitrate. *J. Fish. Res. Bd. Can.*, **28**, 1763–82.

Schneider, S. H. (1989) *Global Warming*. Sierra Club Books, San Francisco, California. 317 pp.

Schowalter, T. D., Hargrove, W. W. and Crossley, Jr., D. A. (1986) Herbivory in forested ecosystems. *Annu. Rev. Entomol.*, **31**, 177–96.

Schowalter, T. D., Webb, J. W. and Crossley, Jr., D. A. (1981) Community structure and nutrient content of canopy arthropods in clearcut and uncut forest systems. *Ecology*, **62**, 1010–1019.

Schultz, A. M. (1964) The nutrient recovery hypothesis for arctic microtine cycles. II. Ecosystem variables in relation to arctic microtine cycles. In *Grazing in Terrestrial and Marine Environments: A Symposium of the British Ecological*

Society, Bangor, 11–14 April 1962 (ed. D. J. Crisp), pp. 57–68. Blackwell Scientific Publications, Oxford.

Schultz, A. M. (1969) A study of an ecosystem: The Arctic tundra. In *The Ecosystem Concept in Natural Resource Management* (ed. G. Van Dyne), pp. 77–93. Academic Press, New York.

Seadler, A. W. and Koonce, J. F. (1976) Phosphorus remineralization by the bluegill sunfish (*Lepomis macrochirus*). Annual Meeting of the Ecological Society of America, New Orleans.

Seale, D. B. (1980) Influence of amphibian larvae on primary production, nutrient flux, and competition in a pond ecosystem. *Ecology*, **61**, 1531–50.

Seastedt, T. R., Crossley, D. A. and Hargrove, W. W. (1983) The effects of low-level consumption by canopy arthropods on the growth and nutrient dynamics of black locust and red maple trees in the southern Appalachians. *Ecology*, **64**, 1040–8.

Sedlacek, J. D., Barrett, G. W. and Shaw, D. R. (1988) Effects of nutrient enrichment on the Auchenorrhyncha (Homoptera) in contrasting grassland communities. *J. Appl. Ecol.*, **25**, 537–50.

Shapiro, J. (1980) The need for more biology in lake restoration. Com. No. 183, Limnology Research Centre University of Minnesota (Mimeo).

Shapiro, J., Forsberg, B., Lamarra, V., Linkmark, G., Lynch, M., Smeltzer, E. and Zoto, G. (1982) Experiments and experiences in biomanipulation – studies of biological ways to reduce algal abundance and eliminate blue–greens. EPA-600/3-82-096. Corvallis, Oregon. Corvallis Environmental Research Laboratory. U.S. Environmental Protection Agency.

Shapiro, J. and Wright, D. I. (1984) Lake restoration by biomanipulation: Round Lake, Minnesota, the first two years. *Freshw. Biol.*, **14**, 371–83.

Shaver, G. R. and Chapin, III, F. S. (1980) Response to fertilization of various growth forms in an Alaska tundra: nutrient accumulation and growth. *Ecology*, **61** 662–75.

Shaver, G. R. and Melillo, J. M. (1984) Nutrient budgets of marsh plants: efficiency concepts and relation to availability. *Ecology*, **65**, 1491–510.

Shelford, V. E. (1913) *Animal Communities in Temperate America*. University of Chicago Press, Chicago.

Sheppard, C. W. (1962) *Basic Principles of the Tracer Method*. John Wiley and Sons, New York.

Short, R. A. and Maslin, P. E. (1977) Processing of leaf litter by a stream detritivore: Effect on nutrient availability to collectors. *Ecology*, **58**, 935–8.

Shugart, H. H. (1984) *A Theory of Forest Dynamics*. Springer-Verlag, New York. 278 pp.

Simenstad, C. A., Estes, J. A. and Kenyon, K. W. (1978) Aleuts, sea otters, and alternate stable-state communities. *Science*, **200**, 403–11.

Sjoberg, S. (1977) Are pelagic systems inherently unstable? A model study. *Ecol. Modell.*, **3**, 17–37.

Slatkin, M. (1974) Competition and regional coexistence. *Ecology*, **55**, 128–34.

Smith, V. H. (1979) Nutrient dependence of primary productivity in lakes. *Limnol. Oceanogr.*, **24**, 1051–964.

Smith, S. V. (1984) Phosphorus versus nitrogen limitation in the marine environment. *Limnol. Oceanogr.*, **29**, 1149–60.

Solomon, A. M. (1986) Transient response of forests to CO_2-induced climate change: Simulation modeling experiments in eastern North America. *Oecologia*, **68**, 567–79.

Solomon, B. P. (1983). Compensatory production in *Solannum carolinense* following attack by a host specific herbivore. *J. Ecol.*, **71**, 681–90.

Sommer, U. (1984) The paradox of the plankton: Fluctuations of phosphorus availability maintain diversity of phytoplankton in flow-through cultures. *Limnol. Oceanogr.*, **29**, 633–6.

Sprent, J. I. (1987) *The Ecology of the Nitrogen Cycle. Cambridge Studies in Ecology.* Cambridge University Press, Cambridge. 151 pp.

Stachurski, A. and Zimka, J. R. (1975) Methods of studying forest ecosystems: leaf area, leaf production, and withdrawal of nutrients from leaves of trees. *Ekologia Polska*, **23**, 637–48.

Stark, N. (1973) *Nutrient Cycling in a Jeffrey Pine Ecosystem.* Institute for Microbiology, University of Montana, Missoula.

Steele, J. H. (1962) Environmental control of photosynthesis in the sea. *Limnol. Oceanogr.*, **7**, 137–50.

Steele, J. H. (1974a) *Stability of plankton ecosystems.* In *Ecological Stability* (eds M. B. Usher and M. H. Williams) pp. 179–191. Chapman and Hall Publishers, London.

Steele, J. H. (1974b) *The Structure of Marine Ecosystems.* Harvard University Press, Cambridge, Massachusetts. 128 pp.

Steele, J. H. (ed.) (1978) *Spatial Pattern in Plankton Communities. NATO Conference Series IV: Marine Sciences* Vol. 3. Plenum, New York.

Steele, J. H., and Henderson, E. W. (1981) A simple plankton mode. *Amer. Natur.*, **117**, 676–91.

Steele, J. H. and Frost, B. W. (1977) The structure of plankton communities. *Philos. Trans. R. Soc. London., B., Biol. Sci.*, **280**, 485–534.

Stenseth, N. C. (1978) Do grazers maximize individual plant fitness? *Oikos*, **31**, 299–306.

Sterner, R. W. (1986) Herbivore's direct and indirect effects on algal populations. *Science*, **231**, 605–7.

Strain, B. and Cure, J. (1985) Direct effects of increasing carbon dioxide on vegetation. (DOE/ER-0236). U.S. Department of Energy, Washington, D.C. Available from NITS, Springfield, Virginia.

Strong, D. R., Lawton, J. H. and Southwood, T. R. E. (1984) *Insects on Plants: Community Patterns and Mechanisms.* Blackwell, Oxford.

Stumm, W. and Morgan, J. J. (1981) *Aquatic Chemistry*, 2nd edn, John Wiley and Sons, New York. 780 pp.

Suttle, C. A., Stockner, J. G., Shortreed, K. S. and Harrison, P. J. (1988) Time-courses of size-fractionated phosphate uptake: Are larger cells better competitors for pulses of phosphate than smaller cells? *Oecologia (Berlin)*, **74**, 571–6.

Swift, M. J., Heal, O. W. and Anderson, J. M. (1979) *Decomposition in Terrestrial Ecosystems. Studies in Ecology* Vol. 5. University of California Press, Berkeley. 372 pp.

Swift, M. J., Russell-Smith, A. and Perfect, T. J. (1981) Decomposition and mineral-nutrient dynamics of plant litter in a regenerating bush-fallow in sub-humid tropical Nigeria. *J. Ecol.*, **69**, 981–95.

Switzer, G. L. and Nelson, L. E. (1972) Nutrient accumulation and cycling in loblolly pine (*Pinus taeda* L.). Plantation ecosystems: The first twenty years. *Soil Sci. Am. Proc.*, **36**, 143–7.

Tanner, E. V. J. (1985) Jamaican montane forests: Nutrient capital and cost of growth. *J. Ecol.*, **73**, 553–68.

Tast, J. and Kalela, O. (1971) Comparison between rodent cycles and plant production in Finnish Lapland. *Ann. Acad. Sci. Fenn., Ser. A, IV*, **186**, 1–14.

Taube, M. (1985) *Evolution of Matter and Energy on a Cosmic and Planetary Scale.* Springer-Verlag, New York. 289 pp.

Taylor, R. J. (1984) *Predation.* Chapman and Hall, New York. 166 pp.

Thomann, R. V., Winfield, R. P. and DiToro, D. M. (1974) Modeling of phytoplankton in Lake Ontario. *Proc. Conf. Great Lakes Res., 17th Int. Assoc. Great Lakes Res.*, pp. 135–49.

Tilman, D. (1977) Resource competition between planktonic algae: An experimental and theoretical approach. *Ecology*, **58**, 338–48. International Association for Great Lakes Research, Madison, Wisconsin.

Tilman, D. (1982) *Resource Competition and Community Structure. Monographs in Population Biology.* Princeton University Press, Princeton, New Jersey. 296 pp.

Tilman, D. and Kilham, S. (1976) Phosphate and silicate growth and uptake kinetics of the diatoms *Asterionella formosa* and *Cyclotella meneghiniana* in batch and semicontinuous culture. *J. Phycol.*, **12**, 375–83.

Tilman, D., Mattson, M. and Langer, S. (1981) Competition and nutrient kinetics along a temperature gradient: An experimental test of a mechanistic approach to niche theory. *Limnol. Oceanogr.*, **26**, 1020–33.

Titman, D. (1976) Ecological competition between algae: Experimental confirmation of resource-based competition theory. *Science*, **192**, 463–5.

Toggweiler, J. R., Sarmiento, J. L., Najjar R. and Papademetriou, D. (1987) *Models of Chemical Cycling in the Ocean.* Ocean Tracers Laboratory Technical Report No. 4. Atmospheric and Oceanic Sciences Program, Department of Geological and Geophysical Sciences, Princeton University, Princeton, New Jersey.

Trabalka, J. R., Edmonds, J. A., Reilly, J. M., Gardner, R. H. and Voorhees, L. D. (1985) Human alterations of the global carbon cycle and the projected future. In *Atmospheric Carbon Dioxide and the Global Carbon Cycle* (ed. J. R. Trabalka), pp. 247–288. DOE/ER-0239. U.S. Government Printing Office, Washington D.C.

Tritton, L. M., Martin, C. W., Hornbeck, J. W. and Pierce, R. S. (1987) Biomass and nutrient removals from commercial thinning and whole-tree clearcutting of central hardwoods. *Environ. Manag.*, **11**, 659–66.

Trudinger, P. A. and Swaine, D. J. (eds) (1979) *Biogeochemical Cycling of Mineral-Forming Elements. Studies in Environmental Science 3.* Elsevier, Amsterdam.

Turner, J., Cole, D. W. and Gessel, S. P. (1976) Mineral nutrient accumulation and cycling in a stand of red alder (*Alnus rubra*). *J. Ecol.*, **64**, 965–74.

Ulanowicz, R. E. (1986) *Growth and Development: Ecosystems Phenomenology.* Springer-Verlag, New York. 203 pp.

Van Cleve, K., Dyrness, C. T., Viereck, L. A., Fox, J., Chapin, III, F. S. and Oechel, W. (1983) Taiga ecosystems in interior Alaska. *BioScience*, **33**, 39–44.

Vandermeer, J. (1980) Indirect mutualism: Variations on a theme by Stephen Levine. *Am. Nat.*, **116**, 441–8.

Van Dyne, G. M. (1966) *Ecosystems, Systems Ecology, and Systems Ecologists.* ORNL 3957. Oak Ridge National Laboratory, Oak Ridge, Tennessee. 40 pp.

Van Dyne, G. M. (1969a) *Grasslands Management, Reseach, and Training Viewed in a Systems Context.* Range Science Department, Science Series No. 3, Colorado State University, Fort Collins, Colorado.

Van Dyne, G. M. (1969b) *Ecosystem Concept in Natural Resource Management.* Academic Press, New York. 383 pp.

Vanni, M. J. (1987) Effects of nutrients and zooplankton size on structure of a phytoplankton community. *Ecology*, **68**, 624–35.

Varga, R. S. (1960) *Matrix Iterative Analysis.* Prentice-Hall, New York.

Verkaar, H. J. (1986) When does grazing benefit plants? *Trends Ecol. Evol.*, **1**, 168–9.

Vickery, P. J. (1972) Grazing and net primary production of a temperate grassland. *J. Appl. Ecol.*, **9**, 307–14.

Visser, S. (1986) The role of the soil invertebrates in determining the composition of soil microbial communities. In *Ecological Interactions in the Soil Environment. Plants, Microbes and Animals* (ed. A. H. Fitter). Blackwell, Oxford.

Vitousek, P. (1982) Nutrient cycling and nutrient use efficiency. *Am. Nat.*, **119**, 553–72.

Vitousek, P. M. and Melillo, J. M. (1979) Nitrate losses from disturbed forests: Patterns and mechanisms. *For. Sci.*, **25**, 605–19.

Vitousek, P. M. and Reiners, W. A. (1975) Ecosystem succession and nutrient retention: a hypothesis. *BioScience*, **25**, 376–81.

Voinov, A. A. and Svirizhev, Yu. M. (1984) A minimal model of eutrophication in freshwater systems. *Ecol. Modell.*, **23**, 277–92.

Vollenweider, R. A. (1969) Moglichkeiten und Grenzen elementar Modelle der Stoffbilanz von Seen. *Arch. Hydrobiol.*, **66**, 1–36.

Volterra, V. (1931) *Theorie Mathematique de la Luute pour la Vie*. Gauthier-Villars, Paris.

Walsh, J. J. (1975) A spatial simulation model of the Peru upwelling ecosystem. *Deep-Sea Res.*, **22**, 201–36.

Walters, C. J., Krause, E., Neill, W. E. and Northcote, T. G. (1987) Equilibrium models for seasonal dynamics of plankton biomass in four oligotrophic lakes. *Can. J. Fish. Aquati. Sci.*, **44**, 1002–17.

Waring, R. H. and Schlesinger, W. H. (1985) *Forest Ecosystems: Concepts and Management*. Academic Press, New York. 340 pp.

Watt, A. S. (1947) Pattern and process in a plant community. *J. Ecol.*, **43**, 490–506.

Webster, J. R. (1975) Analysis of potassium and calcium dynamics in stream ecosystems in three Southern Appalachian watersheds of contrasting vegetation. Ph.D. Dissertation, University of Georgia, Athens, Georgia.

Webster, J. R. and Patten, B. C. (1979) Effects of watershed perturbation on stream potassium and calcium dynamics. *Ecol. Monogr.*, **19**, 51–72.

Webster, J. R., Waide, J. B. and Patten, B. C. (1975) Nutrient cycling and stability of ecosystems. In *Mineral Cycling in Southeastern Ecosystems*. (eds F. G. Howell, J. B. Gentry and M. H. Smith) pp. 1–27. ERDA Symposium Series, Washington D.C.

Weinstein, D. A., Shugart, H. H. and West, D. C. (1982) The long-term retention properties of forest ecosystems: a simulation investigation. ORNL/TM-8472. Oak Ridge National Laboratory, Oak Ridge, Tennessee.

Welch, E. B. (1980) *Ecological Effects of Waste Water*. Cambridge University Press, Cambridge. 337 pp.

Welch, E. B., Hendrey, G. R. and Stoll, R. K. (1975) Nutrient supply and the production and biomass of algae in four Washington lakes. *Oikos*, **26**, 47–54.

Wesley, J. P. (1974) *Ecophysics: The Applications of Physics to Ecology*. Charles C. Thomas, Springfield, Illinois. 340 pp.

West, D. C., Shugart, H. H. and Botkin, D. B. (eds.) (1981) Forest Succession: Concepts and Applications. Springer–Verlag, New York. 517 pp.

Wetzel, R. G. (1983) *Limnology*. Saunders, Philadelphia, Pennsylvania. 743 pp.

White, T. C. R. (1974) A hypothesis to explain outbreaks of looper caterpillars with special reference to populations of *Selidosema suavis* in a plantation of *Pinus radiata* in New Zealand. *Oecologia (Berlin)*, **16**, 279–301.

White, T. C. R. (1978) The importance of a relative shortage of food in animal ecology. *Oecologia (Berlin)*, **3**, 71–86.

Whittaker, R. H. (1961) Experiments with radiophosphorus tracer in aquarium microcosms. *Ecol. Minogr.*, **31**, 157–8.

Wielgolaski, F. E. (1975) *Fennoscandian Tundra Ecosystems*. Springer-Verlag, New York.

Williamson, S. C. (1983) The herbivore optimization hypothesis: A sampling design, experimental examination, and simulation. Ph.D. Thesis, Colorado State University, Fort Collins, Colorado. 120 pp.

Witkamp, M. (1971) Soils as components of ecosystems. *Annu. Rev. Ecol. System.*, **2**, 85–110.

Woodin, S. A. and Yorke, J. A. (1974) Disturbance, fluctuating rates of resource recruitment, and increased diversity. In *Ecosystem Analysis and Prediction. Proceedings of a Conference on Ecosystems, Alta, Utah* (ed. S. A. Levin), pp. 38–41. SIAM, Philadelphia, Pennsylvania.

Woodmansee, R. G. (1978) Additions and losses of nitrogen in grassland ecosystems. *BioScience*, **28**, 448–53.

Woodwell, G. M., Hobbie, J. E., Houghton, R. A., Melillo, J. M., Moore, B., Peterson, B. J. and Shaver, G. R. (1983) Global deforestation: Contribution to atmospheric carbon dioxide. *Science*, **222**, 1081–6.

Wright, H. E., Jr. (1974) Landscape development, forest fires, and wilderness management. *Science*, **186**, 487–95.

Wroblewski, J. S., Sarmiento, J. L. and Flierl, G. R. (1988) An ocean basin scale model of plankton dynamics in the North Atlantic. 1. Solutions for the climatological conditions in May. *Global. Geochem. Cycles*, **2**, 199–218.

Wulff, F. (1989) A time-dependent budget model for nutrients in the Baltic Sea. *Glob. Biogeochem. Cycles*, **3**, 63–78.

Yodzis, P. P. (1981) The stability of real ecosystems. *Nature*, **289**, 674–6.

Yodzis, P. P (1984) Energy flow and the vertical structure of real ecosystems. *Oecologia*, **65**, 86–8.

Zaret, T. M. and Paine, R. T. (1972) Species introduction in a tropical lake. *Science*, **183**, 449–55.

Zeeman, E. C. (1972) Differential equations for the heartbeat and nerve impulse. In *Towards a Theoretical Biology 4: Essays* (ed. C. H. Waddington) Aldine-Atherton, Chicago. pp. 8–67.

Index

All references in *italics* represent figures, those in **bold** represent tables.